Space and Society

The Space and Society series explores a broad range of topics in astronomy and the space sciences from the perspectives of the social sciences, humanities, and the arts. As humankind gains an increasingly sophisticated understanding of the structure and evolution of the universe, critical issues arise about the societal implications of this new knowledge. Similarly, as we conduct ever more ambitious missions into space, questions arise about the meaning and significance of our exploration of the solar system and beyond. These and related issues are addressed in books published in this series. Our authors and contributors include scholars from disciplines including but not limited to anthropology, architecture, art, environmental studies, ethics, history, law, literature, philosophy, psychology, religious studies, and sociology. To foster a constructive dialogue between these researchers and the scientists and engineers who seek to understand and explore humankind's cosmic context, the Space and Society series publishes work that is relevant to those engaged in astronomy and the space sciences, while also being of interest to scholars from the author's primary discipline. For example, a book on the anthropology of space exploration in this series benefits individuals and organizations responsible for space missions, while also providing insights of interest to anthropologists. The monographs and edited volumes in the series are academic works that target interdisciplinary professional or scholarly audiences. Space enthusiasts with basic background knowledge will also find works accessible to them.

More information about this series at http://www.springer.com/series/11929

Mark Lupisella

Cosmological Theories of Value

Science, Philosophy, and Meaning in Cosmic Evolution

Springer

Mark Lupisella
The Horizons Project
Ashburn, United States

ISSN 2199-3882 ISSN 2199-3890 (electronic)
Space and Society
ISBN 978-3-030-25337-0 ISBN 978-3-030-25339-4 (eBook)
https://doi.org/10.1007/978-3-030-25339-4

Cover design: Paul Duffield

This Springer imprint is published by the registered company Springer Nature Switzerland AG
The registered company address is: Gewerbestrasse 11, 6330 Cham, Switzerland

To my daughter Emilee, and to those yet to be.

Enjoy The Show…

Foreword by Steven J. Dick

The human search for meaning in our expansive universe has a long and checkered history, particularly in the realms of religion, ethics, and philosophy. Attempts to anchor that meaning in the findings of science have a shorter and even more controversial history. Three decades ago the renowned historian of science John Greene, for example, cautioned in no uncertain terms that naturalism—the idea that ethical, religious, and social beliefs should be based on our knowledge of the natural world rather than supernatural speculations—is just another worldview, no more privileged than supernaturalism or anything else to serve as the source of value and meaning in human life. He pointed out that scientists such as Ernst Haeckel, Julian Huxley, Ralph Burhoe, and E. O. Wilson "all typify the scientist-ideologue bent on saving society by promulgating new ethics and a new religion claiming the sanction of science." The pattern is "depressingly familiar," in Greene's view: "When will scientists and others learn that naturalism is a philosophical point of view with no more claim to the status of science than any other philosophical viewpoint, whether Marxian, Freudian, Russellian, Whiteheadian, or whatever. Scientists have as good a right to expound their philosophical, ethical and religious views as anyone else, but they have no right to palm these off as the findings of science" (Greene 1989, 404). The attempt to derive meaning and morals from what we know about the universe even has a derogatory name: "the naturalistic fallacy," and precautions against inferring value from fact, "ought" from "is," go back at least to the philosopher David Hume in the eighteenth century.

In my view, Greene's attitude is both instructive and dead wrong. It is instructive in the sense that science-oriented values are far from the sole source of meaning and value in life. Or as the author puts it in Chap. 3 of this provocative volume, facts should certainly inform but not completely determine values. And it is dead wrong in the sense that science has *at least as much* claim to inform ethics and theology as any other worldview, and perhaps more, since being grounded in the natural world gives it a pillar of support that supernatural religions do not have, numerous attempts at natural theology notwithstanding. Exactly, how to draw meaning from the findings of science is, of course, open to interpretation and argument. In this volume, scientist–philosopher Mark Lupisella—fully aware of the naturalistic fallacy and its nuances—describes a series of worldviews that can be characterized as encompassing a morally creative cultural cosmos, a universe driven largely by moral and creative pursuits. In other words, a worldview in which meaning and value may be bootstrapped from the universe.

What does the hard, cold, but beautiful and wondrous cosmos have to do with *us*, in particular with the subjective idea of value? In this volume, the author places cosmic evolution at front and center stage, arguing that human destiny is intimately connected to the cosmos and cosmic evolution. He sets forth three cosmological theories of value, both in the sense of "value" as basic principles that motivate actions and as ideas that are valuable for individuals and cultures. In its simplest form that he terms "cosmic reverence," we can certainly value our place in 13.8 billion years of cosmic evolution—something that seems simple but that requires a certain cosmic consciousness. We can go one more step and argue that we are coevolving with the universe—that the universe gains some value from us in a two-way relationship, a kind of cosmocultural coevolution. Or we can invoke what Lupisella calls the "connection-action principle," the idea that connectedness is an intrinsic property of the universe manifested in relations and actions that account for the dynamism and creativity of cosmic evolution, including life and intelligence. This property is consistent with what we observe in terms of the diversity of relations all around us, whether process philosophy or quantum mechanics, but raises questions of determinism and free will. Nevertheless, Lupisella argues that such a dynamic creative universe can serve as the basis for cosmological intrinsic value, the basis for a kind of cosmocentric ethic.

Ideas of religion, ethics, spirituality, and meaning grounded in nature rather than the supernatural are commonplace. Books such as Ursula Goodenough's *The Sacred Depths of Nature* (1998) or Brian Swimme and Thomas Berry's *The Universe Story* (1992) spring to mind, and Lupisella discusses these and other precedents in this volume. More substantially, Clement Vidal's book *The Beginning and the End: The Meaning of Life in a Cosmological Perspective* (2014) provides another provocative view of how the cosmos may have meaning for humans and intelligent life. But with his three cosmological theories of value, Lupisella goes well beyond the bounds of most books on naturalism and into fundamental questions about the nature of the universe and our relation to it.

What are the practical implications of what might seem like academic deliberations? For starters, numerous theories of moral status vie for human attention, ranging from the anthropocentric to the biocentric, planetocentric, and cosmocentric. Which theory of moral status we choose as individuals, cultures, and as a species will determine how we treat our fellow humans and all life on Earth. Moreover, choosing an anthropocentric as opposed to a cosmocentric ethic in interacting with any extraterrestrial intelligence may have disastrous consequences. Even if we never encounter intelligent life beyond the Earth (and I am betting we will), placing life and intelligence in a cosmic context is an important endeavor. Deriving meaning from a cosmic context is even more important, critics notwithstanding. And it is critical to discuss these ideas now, anticipating the day when we do discover life beyond Earth. Fortune in the face of unexpected discoveries favors the prepared mind, as Louis Pasteur reminded us, improving on the ancient Roman proverb that fortune favors the bold, the brave, or the strong.

In pursuit of its goals, this volume ranges from the sciences of cosmic evolution, relativity, and quantum mechanics, to value theory and process philosophy, all with the goal of exploring how they may relate to *you* and humanity in the sense of worldviews, ethics, and meaning. You can decide whether your worldview encompasses simply a reverence for the cosmos, whether you think we are coevolving with that cosmos, or whether you wish to travel with the author to the realm of the dynamic creative universe that has resulted in us and other intelligence, perhaps a universe, as physicist Freeman Dyson evocatively put it, that "knew we were coming." To read Lupisella is to have a mind-boggling experience, to want to race to references, to want to know more. And arguably that process—your decisions and mine—is all part of the coevolution of cosmos and culture.

Ashburn, United States

Steve J. Dick
Former Baruch S. Blumberg NASA
Library of Congress Chair in Astrobiology
Former NASA Chief Historian

Reference

Greene, J. C. (1989). "Afterword" to James R. Moore, ed., *History, humanity and evolution: Essays for John C. Greene* (pp. 403–413: 404). Cambridge: Cambridge University Press.

Preface

'Theory' derives from the Greek 'theoria', which has the same root as 'theatre', namely 'theaomai', which means 'to view'.

In this spirit let us by all means 'enjoy the show'...

—Paavo Pylkkänen[1]

Cosmic evolution, with its wondrous creations and mind-boggling science, arouses. Like any good show, it stirs us, provokes us to wonder, and inspires us with awe. Cosmic evolution calls us to explore some of our biggest questions.

Indeed, the older I get the more sensitive I am to the apparent conundrum that so many of us are in this cosmic show without really knowing why. We have some idea as to *how*, and sometimes *how* can help address *why*, but *why* can be a very different kind of question, often implying some meaning or purpose or intent. Sure, our curious overactive minds could be playing tricks on us, relentlessly knocking on unanswerable and even misguided questions. Or, I, like many others, may be mistakenly averse to traditional religious answers that tend to invoke omniscient, omnipotent, and often perplexing Gods.

Thinking too much about big questions and constructing grand "cosmic" worldviews can certainly lead us astray and may come across as too "new agey" for some.[2] Such pursuits may merely reflect psychological predispositions to find patterns, meaning, and purpose where there may be none. Pursuits to "understand it all" when the truth ultimately eludes us can be deeply problematic and sometimes dangerous. But it is difficult to know if and how adventures of our minds will be helpful or misguided. It is essentially impossible to know in advance where scientific and philosophical explorations will lead, so instead of paralysis in the face of uncertainty, we embark on endless knowledge-seeking and contemplation. Indeed, many of us cannot help but engage in the often delightful and sometimes unsettling

[1]Pylkkänen, Paavo T. I. 2007. *Mind, Matter, and The Implicate Order*. The Frontiers Collection. Springer, Berlin, Heidelberg. P. 157.

[2]Ted Peters explored New Age movements in his 1991 book notably titled, *The Cosmic Self: A Penetrating Look at Today's New Age Movements* (Harper Collins).

pursuits of meaning and purpose—of human conjecture that has occupied and pleasantly surprised humanity for millennia.

This book is a reflection of such an endeavor. It is a collection of thoughts and writing[3] from many years of exploration that can be traced back to early childhood (and maybe back much further if one wants to blame genetics!). From my early days as a questioning Catholic altar boy in Sunday school, I found many responses wanting and confusing, if not irrational, unjustifiable, and misleading. I have a vivid memory of exceeding a Sunday school teacher's threshold for questions about the Trinity when she asked the priest to come in to class to talk to me. He leaned over me and gave a refreshingly honest and understandably simple response in front of the class: "Well young man, ultimately it's a leap of faith."

I knew exactly what he meant, and I appreciated his response. I sensed he had been through this kind of thing before. He knew what kind of response to give. He knew he had the difficult task of intellectually justifying problematic details like the Holy Trinity to a persistently inquisitive and potentially uncooperative youngster—in front of an entire class, no less—for which there were almost certainly other similarly perplexed students. He was not going to argue the point. He was not going to explain something that seemed unexplainable. He was not going to offer logic, reason, or other intellectual justification for ideas that seemed to defy such things.

Instead he offered faith. He knew a truth of sorts for himself and others and maybe a broader truth about faith: That it is a way out and maybe some kind of way in. It is at least a window, or a release valve, or maybe an escape route. It is a window to see in and out of, to climb in and out of. Faith can be a way out of having to accept uncertainties, a way out of having to reserve judgment about ultimate questions, and a way out of having to do the hard work of finding more challenging explanations that might be independent of human desires, human angst, and human storytelling. But the window of faith can also be a way in. It can be a way to understand ourselves and much else we do not understand. Faith can be a way to see, to point out and call out—a way to understand what we deeply need, what we deeply desire, what we deeply love.

[3]Much if this book is based on previous work by the author: (1) Lupisella (2016), Cosmological Theories of Value: Relationalism and Connectedness as Foundations for Cosmic Creativity, in *The Ethics of Space Exploration*, Tony Milligan and James Schwartz (Eds), Springer. (2) Lupisella (2015), Life, Intelligence, and the Pursuit of Value in Cosmic Evolution, in *The Impact of Discovering Life Beyond Earth*, ed. Steven. J. Dick. Cambridge University Press. (3) Lupisella (2009) Cosmocultural Evolution: The Coevolution of Cosmos and Culture and the Creation of Cosmic Value. In *Cosmos and Culture: Cultural Evolution in a Cosmic Context*, edited by Steven J. Dick and Mark L. Lupisella, 321–359. NASA SP-2009-4802. http://history.nasa.gov/SP-4802.pdf, corrected version here: https://innovim.academia.edu/MarkLupisella, (4) Lupisella (2009), The Search for Extraterrestria Life: Epistemology, Ethics and Worldviews, in *Exploring the Origin, Extent, and Future of Life*, ed. Constance Bertka, American Association for the Advancement of Science, Cambridge University Press. (5) Lupisella and Logsdon (1997), "Do We Need a Cosmocentric Ethic?" Paper IAA-97-IAA.9.2.09 presented at the International Astronautical Federation Congress. American Institute of Aeronautics and Astronautics, Turin.

I could have responded to our priest by saying how faith often seems groundless or baseless or misleading—how it seems unjustified for just about anything. But I knew he was saying that it is essentially a choice—a choice to leap. But does this choice ask too much? Our priest offered a choice to essentially turn way from reason, logic, and rational thought—from the cold, calculated, relentless, complex, and often inconvenient, sometimes inadequate and unsatisfying, pounding hammer of reason.

But then, a logic-seeking mind in search of sound reasoning and rational thought in all matters of human experience may ask too much as well. The human story can often seem infinitely complex and riddled with profound and fundamental uncertainties—for which no truly "good" explanations exist, for which no philosophical explorations help, and for which no adventure can satisfy. So our philosophical pursuits can and should include a sense of responsibility and balance for a careful examination of human emotions, evidence, and reasoning, along with open-minded contemplation—and yes, a certain amount of speculation that can often be far-reaching and uncomfortable.

Practical intellectual utility, particularly for large numbers of people, is important of course, but it need not be our sole motivation—it need not be our sole end in investigating our experiences and the wider world. We can also be motivated by the charm and inspiration of potentially endless philosophical adventures. It is hoped this book does some of both—provide some practical utility as well as some philosophical inspiration and philosophical fun for its own sake. Perhaps the best knowledge-seeking does both. Perhaps the best philosophy does both. Perhaps wisdom results from both.

So maybe after all there is a little faith here, a little faith in reason, a little faith in science, a little faith in philosophy, a little faith in others, and maybe even a little faith in faith. Perhaps we can have a little faith in simply taking the journey—alone and together. But faith in these things is a very different kind of faith—if it is any kind of faith at all. Our journey does not need the kind of leap of faith that is often called for, or at least implied, by many religions and other domains of human experience. Our journeys can be grounded, incremental, evolutionary, and reasoned explorations. Our explorations do not have to invoke faith without some kind of support or evidence or reason. We do not have to rely on "blind faith" or any one orientation to guide us. Indeed, humanity has a long history of making great strides through a plurality of pursuits, including forms of faith, emotional and psychological motivations, scientific pursuits, and philosophical exploration.

In that spirit then, this book is an exploration of a plurality of ideas and theories, and of many avenues and methods—a kind of scientific and philosophical buffet from which we might cobble together enticing worldviews. This is not quite the philosophical or scientific treatment that either academic community might normally require. There is admittedly a healthy dose of "fence-sitting" here—something I learned about the hard way in Mr. Kemmer's high school social studies class when I didn't take a position on a minor school issue. Indeed, I ask for the reader's patience with the overuse of noncommital words and abundance of repetition, rephrasing and hesitation as we poke at squishy nuances and explore potentially unanswerable questions. I ask for the reader's broadest "open-minded skepticism" as we zig-zag a bit loosely among diverse and often incomplete concepts, attempting to see things from different sides of rickety fences.

In that light, this book can be seen as a kind of rough map to help navigate and romance the universe, to perhaps see things a little differently. And if some satisfaction, some inspiration, and some joy lie in the journey, then let us look to our philosophical adventures as ends unto themselves. Let us seek, embrace, and share our explorations not only because they are their own reward, but also because we are on this journey together.

Ashburn, United States

Acknowledgements

There are many to thank for this journey. Thanks to Doug Vakoch as the series editor and patient guide, as well as others at Springer such as Ramon Khanna who provided feedback and helped with the publishing process, including obtaining helpful anonymous reviewer feedback. Thanks to others such as Kokila Durairaj and the Springer team for shepherding this book through production.

Thanks to those who have helped me and inspired me, if even only sometimes through fleeting but impactful interactions—their conversations and compelling writing have been a kind fuel: Iván Almár, Jacques Arnould, Cynda Arsenault, Béla Bánáthy, Connie Barlow, Mike Berger, Connie Bertka, John Billingham, Linda Billings, Susan Blackmore, Howard Bloom, Baruch Blumberg, Larry Bohlen, Chris Boshuizen, Penny Boston, Stuart Brand, Roshan Chaddha, Eric Chaisson, David Christian, Chris Chyba, Carol Cleland, Ian Crawford, Erin Daly, Jim Dator, Paul Davies, Kathryn Denning, Steve Dick, Freeman Dyson, Pascale Ehrenfreund, Martyn Fogg, David Fromkin, Stuart Gill, Tom Gleason, Jerry Glenn, Brian Green, Ed Greville, Gene Hargrove, Tristan Harris, Al Harrison, Robert Haynes, Gerda Horneck, Bruce Jakosky, Sarbmeet Kanwal, Ramon Khanna, John Logsdon, Alan Marshall, Will Marshall, John Mather, Alex Mazarr, Chris Mckay, Wendell Mendell, Jake Metcalf, Michael Michaud, Tony Milligan, Ted Peters, Zach Pirtle, Margaret Race, Kim Stanley Robinson, Holmes Rolston, Jessy Schingler, Jim Schwartz, George Smith, Kelly Smith, Jill Tarter, Allen Tough, Clément Vidal, Rich Vondrak, Edward O. Wilson, Robert Zubrin.

Thanks to many teachers and professors: Ken Sebens from the Department of Biology at the University of Maryland patiently guided me through my dissertation, "A Theoretical Microbial Contamination Model for a Human Mars Mission," within the interdisciplinary program on Behavior, Ecology, Evolution and Systematics led by Gerald Wilkinson who introduced me to the formal study of animal behavior. Fred Suppe in the University of Maryland Philosophy Department advised me on my Thesis, "Using Artificial Life to Assess the Typicality of Terrestrial Life," and Stephen Brush, a historian who co-chaired The Committee on the History and Philosophy of Science with Fred Suppe taught mind-expanding seminars, leaving indelible impressions—along with many others in the Philosophy

Department such as Michael Devitt, Lindley Darden, Jerrold Levinson, Patricia Greenspan, and Georges Rey. Other professors at Maryland such as John Anderson from the Engineering Department, Richard Hornyak from Physics, and English professor Cathy Barks taught deeply and inspired. Mr. Eggleston and Mr. Kemmer from Hamburg High were engaging and pushed well beyond their curricula boundaries. Mr. Fentzke, my fourth-grade science teacher from Boston Valley, made science memorable and fun.

Thanks to my parents who managed to raise an inquisitive (and challenging) geek. And thanks to my wife Susan and daughter Emilee for rolling through the many family impacts of writing a book.

Contents

About the Author

Mark Lupisella, Ph.D., has a doctorate in evolutionary biology (Program in Behavior, Ecology, Evolution, and Systematics), with his dissertation on "A Theoretical Microbial Contamination Model for a Human Mars Mission." He also has a B.S. in Physics, and a M.A. in Philosophy of Science with a thesis on Using Artificial Life to Assess the Typicality of Terrestrial Life), all from the University of Maryland at College Park. He founded The Horizons Project which aims to improve humanity's ability to address long-term survival challenges. He has worked for NASA for 30 years and presently serves as the exploration research and development manager at the Goddard Space Flight Center. He has worked on Hubble Space Telescope, Human Exploration Programs (with an emphasis on human mission architectures, science integration, and advanced systems), and areas in astrobiology such as planetary protection, artificial life, and societal issues. He assisted the Secure World Foundation in its initial stages of development and also helped lead the development of an ethics committee for Planet Labs (now called Planet). He also serves on the Advisory Council for METI International (Messaging Extraterrestrial Intelligence). He has authored over 35 published works ranging from human Mars missions to the search for extraterrestrial life and space ethics and is the co-editor of the NASA book, *Cosmos and Culture: Cultural Evolution in a Cosmic Context* with previous NASA Chief Historian, Steven Dick.

Chapter 1
Overview and Road Map

As life and intelligence continue to evolve in the universe, it is intriguing to wonder where it all may lead and what, if anything, it may mean. Such contemplations have consumed, inspired, and befuddled human beings since their first stirrings. From philosophy to science, from religion to spirituality, we have wondered about meaning and purpose—in our own lives and that of our wider world. Our wonder and seeking is often captured in broader contexts or "worldviews" that act as frames of reference to help guide us and make sense of the often bewildering complexities of life. Exploring "cosmological worldviews" and any potential implications for life and intelligence in cosmic evolution should be seen in this larger experiential, historical, and cultural context, while recognizing that understanding macroscale social dynamics and cultural evolution can be a complex and tricky pursuit (Denning 2009).

Indeed, cosmological contemplations, particularly those that explore life and intelligence in a cosmic context, often involve wide-ranging speculation and perhaps a certain level of hubris, and perhaps even misguided predispositions and problematic motivations that ultimately justify deep skepticism of such pursuits. But in trying to more thoroughly explore cosmological worldviews and any theories of value that may result, it seems reasonable to explore some specific lines of thinking that place life and intelligence in a cosmic context. Ancient religions, as well as more contemporary religious and spiritual worldviews, have much to say about life and intelligence in this kind of broader "universal" context. Science also has much to say about the past evolution of life and intelligence in a cosmic context—and perhaps even a bit to say about the long-term future. Careful philosophical exploration can leverage science and other forms of knowledge to explore the potential for broader meaning in a cosmological context.

M. Lupisella, *Cosmological Theories of Value*, Space and Society,
https://doi.org/10.1007/978-3-030-25339-4_1

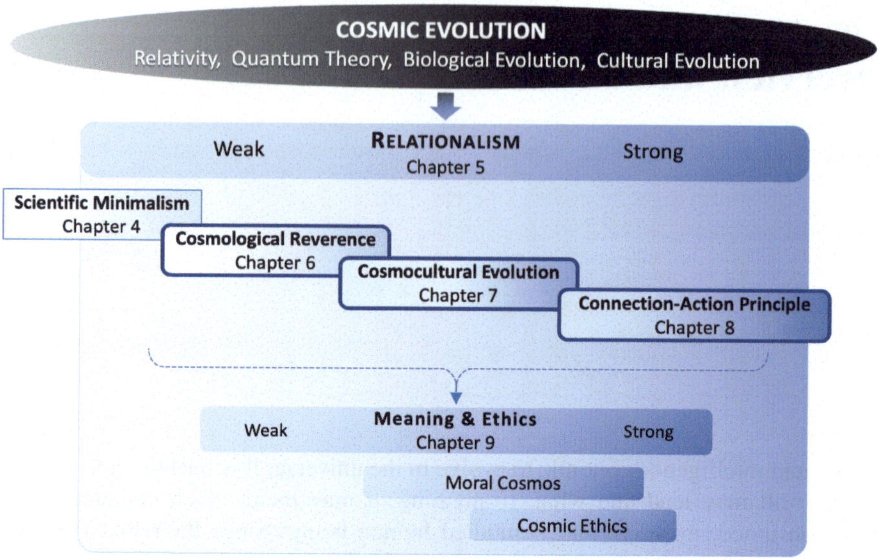

Fig. 1.1 A relationalist framework for cosmological theories of value

It is often understandably suggested that thinking too long term is futile (or at least not the best use of one's time!)[1] Uncertainties, especially those that may be fundamental unknowns, raise legitimate concerns about the utility of this kind of long-term broad-based cosmological speculation. It is hoped that some of those concerns will at least be partially addressed along the way. But for now, let us entertain the answer Stuart Brand gave me when I asked him how far humanity should think into the future: "Infinity," he said.[2]

Figure 1.1 shows an overall landscape, a kind of "road map" and partial chapter map of the book, emphasizing the broad "relationalist framework" that will inform much of the book. While there are exceptions, moving from top to bottom generally takes us from scientific and philosophical foundations (or what might also be considered "ontological" or "metaphysical" foundations) in the upper half of the diagram, to potential philosophical implications for value, meaning, and ethics in the bottom half of the diagram. The darker degrees of shading moving left to right are intended to convey stronger versions of the particular concepts.

[1]Clement Vidal takes this concern head on in his 2014 book, *The Beginning and the End: The Meaning of Life in a Cosmological Perspective.* His preface, provocatively titled "Psychiatry and Cosmological Speculation," unapologetically defends long-term cosmological speculation, and much of his book is devoted to explaining and justifying the philosophical pursuit more generally—emphasizing the careful development of comprehensive and coherent worldviews.

[2]Conversation with Stuart Brand at the NASA/DARPA sponsored 100 Year Spaceship Symposium, October 2, 2011, Orlando, Florida. Vidal (2104) explicitly explores the idea of fully embracing infinity as a value for the most advanced stage of intelligent evolution, especially in the context of a kind of infinite cosmic existence through the endless creation of universes.

The book is divided into two broad parts: Part I, "Foundational Context" (Chaps. 2–5) will briefly touch on some key foundational considerations, and Part II, "Cosmological Worldviews and Implications" (Chaps. 6–9) will develop three specific "cosmological theories of value", where "cosmological" is meant as an adjective to describe the kinds of "theories of value" that can provide theoretical foundations for how we might percieve and "derive" value. And these cosmological theories of value can also be associated with "cosmological worldviews"). We will then further explore some implications for value theory and worldviews more generally, finishing with a final summarizing chapter in Chap. 10.

Part I starts on foundational considerations by exploring some precedent in Chap. 2 of a number of past and present perspectives that relate to life and intelligence in a cosmic context. Chapter 3, "From Replication to Aspiration" then briefly highlights some key aspects of life and intelligence to help set the stage for exploring possible implications and consequences for life and intelligence in a cosmic context. Chapter 3 digs deeper into intelligence and reflects on culture and value theory, and specifically "normative aspiration" and its potential implications for the future of intelligent beings and the universe more broadly.

Chapter 4, on "scientific minimalism," explores various interpretaions of quantum thoery and briefly touches on cosmic evolution, and attempts to makes very few assumptions while examining a narrow set of philosophical implications to help inform a relationalist framework that is developed in Chap. 5. Scientific minimalism is shown in Fig. 1.1 with square corners and without a thick border to help convey that it is not intended to be particularly speculative or the main focus of the book—unlike the three other worldviews that are shown with soft corners and thicker borders, intended to convey they are more speculative areas of focus. Nevertheless, scientific minimalism can be seen as its own kind of worldview and is important for informing the relationalist framework in Chap. 5 and the three main cosmological theories of value and associated worldviews that follow.

Similarly, Chap. 5, which explores a broader relationalist framework, can also be seen as a kind of worldview and will act as a contextual pluralistic philosophical framework for the development of the three specific cosmological worldviews that are developed in Part II in Chaps. 6–8. The broader relational framework in Chap. 5 is introduced after scientific minimalism because scientific minimalism directly informs and leads into the broader relationalist framework. The resulting broader relational framework of Chap. 5 then helps support the development of three cosmological theories of value.

The next three chapters of Part II develop and explore three cosmological theories of value and associated cosmological worldviews. Chapter 6 will explore *cosmological reverence*. Chapter 7 will explore *cosmocultural evolution*. Chapter 8 will explore the *connection-action principle*. Chapter 6 on cosmological reverence is touched on first as a view that ascribes a notable level of value to the universe that results from the scientific view that life and intelligence are intimately related to the universe as a direct product of cosmic evolution. Chapter 7, "Cosmocultural Evolution," goes further and suggests that culture coevolves with the universe in an open-ended but potentially purposeful and meaningful way, influenced heavily by the values and aspirations of valuing agents such as intelligent cultural beings. In Chap. 8, "The Connection-Action Principle," we will wade into philosophical metaphysics and develop the idea that a potentially universal property of connectedness is realized as relations and actions, perhaps providing a conceptual foundation for why and how the universe is dynamic and creative.

While scientific minimalism (Chap. 4) and the relationalist framework (Chap. 5) can both be considered general worldviews in and of themselves, Chaps. 2–5 taken in totality are intended to act primarily as an overall foundation for the three more specific cosmological worldviews developed in Part II in Chaps. 6–8, and also for Chap. 9 that further develops more concrete applications for meaning and ethics. Chapter 2 touches on precedent, and Chap. 3 provides important philosophical context. Chapter 4 explores scientific minimalism and provides important scientific knowledge to help inform subsequent discussions. So, the relationalism overview of Chap. 5 acts as a kind of bridge to take us from the philosophy of Chap. 3 and the science of Chap. 4 to the development of the three specific cosmological worldviews and theories of value in Chaps. 6–8. This can then help us reason through the fact-value and is-ought distinction (closely related to the "naturalistic fallacy") by providing a scientific and philosophical basis for the associated theories of value and cosmic ethics in Chap. 9.

The overall relationalist framework and specific cosmological theories of value will be applied to a number of challenges, with an emphasis on the search for, discovery of, and interaction with potential extraterrestrial life and intelligence. Chapter 10 will further synthesize and summarize, providing an evaluation of the three cosmological worldviews and a tentative conclusion that a "bootstrapped" version of cosmocultural evolution may be the most tenable.

Part I
Scientific and Philosophical Context

Chapter 2
Precedent

Before getting into specific examples of relevant precedent, let us consider some motivational context from Freeman Dyson:

> It is impossible to calculate in detail the long-range future of the universe without including the effects of life and intelligence. It is impossible to calculate the capabilities of life and intelligence without touching, at least peripherally, on philosophical questions. If we are to examine how intelligent life may be able to guide the physical development of the universe for its own purposes, we cannot altogether avoid considering what the values and purposes of intelligent life may be. But as soon as we mention the words value and purpose, we run into one of the most firmly entrenched taboos of twentieth century science.
>
> Hear the voice of Jacques Monod, high priest of scientific rationality, in his book Chance and Necessity: "Any mingling of knowledge with values is unlawful, forbidden."
>
> Monod was one of the seminal minds in the flowering of molecular biology in this century. It takes some courage to defy his anathema. But I will defy him, and encourage others to do so. The taboo against mixing knowledge with values arose during the nineteenth century out of the great battle between the evolutionary biologists led by Thomas Huxley and the churchmen led by Bishop Wilberforce. Huxley won the battle, but a hundred years later Monod and Weinberg were still fighting Bishop Wilberforce's ghost. Physicists today have no reason to be afraid of Wilberforce's ghost. If our analysis of the long-range future leads us to raise questions related to the ultimate meaning and purpose of life, then let us examine these questions boldly and without embarrassment. If our answers to these questions are naive and preliminary, so much the better for the continued vitality of our science.
>
> —Freeman Dyson, "Time Without End," 1979

2.1 Background and Overview

This chapter will briefly touch on a number of worldviews related to life and intelligence in a cosmic context, or what might be called—for lack of a better word—"cosmophilosophy" (Lupisella 2019). The coverage of the views here is not intended as advocacy, but is instead meant to convey some breadth and diversity of a range of

M. Lupisella, *Cosmological Theories of Value*, Space and Society,
https://doi.org/10.1007/978-3-030-25339-4_2

speculative and provocative ideas that can help us see a broad landscape of thinking related to worldviews we will explore in forthcoming chapters.

The breadth of thought that places life and intelligence in cosmic contexts varies widely. We can look to (A) Western and Eastern religious worldviews (e.g., Hinduism), (B) "natural theology" and broader philosophes in the West going back as far as Plato and then to Giordano Bruno in the late 1500s, and Baruch Spinoza in the early 1600s, (C) modern philosophical articulations such as cosmism of the twentieth and twenty-first centuries,[1] Alfred North Whitehead's process philosophy (1929), Lovejoy's treatment of the principle of plenitude (1936), and Teilhard de Chardin's Omega Point Theory (1955), (D) contemporary scientific treatments that place life in a cosmic evolutionary context and contemplate the challenges and implications of the long-term survival of intelligence (e.g., Kardashev 1964; Dyson 1979; Swimme and Berry 1992; Tipler 1994; De Duve 1995; Dick 2000; Chaisson 2005; Davies 2009; Smart 2009; Lupisella 2009; Deutsch 2011; Vidal 2014; Carroll 2016), and (E) the endlessly rich explorations of so much science fiction (e.g., the classic work of Olaf Stapledon, *Star Maker*, in 1937).

The focus in this chapter will be on a few contemporary ideas and thinkers that suggest fairly specific scientifically informed philosophical views of cosmic significance and potential roles for the future life and intelligence in the universe. "Philosophical" is meant here to cast a wide net capturing a variety of views that go beyond narrowly scientific or engineering treatments such as some noted above which generally invoke the more obvious values of survival, increased scientific knowledge, and technological advancement (e.g., as it is tied to energy use (Kardashev 1964) and/or the control of matter more generally). This brief treatment of precedent will then inform the development of a pluralistic philosophical framework that will be explored for additional philosophical and practical implications, such as, among others, the search for, discovery of, and interaction with potential extraterrestrial life and intelligence.

2.2 Cosmological Eschatology[2]

One of the first modern Western nonfiction treatments of the future of life and intelligence in a cosmic context that attempted to weave in the contemporary science of the time comes to us in 1955 from Pierre Teilhard de Chardin in his book, *The*

[1]For a treatment of Russian and American Cosmism, see Albert Harrison, 2013, Russian and American Cosmism: Religion, National Psyche, and Spaceflight. Astropolitics, Vol. 11, Issue 1–2, about which there is a related interview with Harrison by Ross Anderson in The Atlantic, called "The Holy Cosmos: The New Religion of Space Exploration" at: http://www.theatlantic.com/technology/archive/2012/03/the-holy-cosmos-the-new-religion-of-space-exploration/255136/. See also *A Cosmist Manifesto* by Ben Goertzel, 2010, published by Humanity Plus Press.

[2]The meaning of eschatology here is intended to convey a kind of "final state," a slightly more generic use than is often attached to theological worldviews (although theological implications are related to what's covered here).

Phenomenon of Man. In Teilhard's work, we see an application of the current knowledge of evolutionary biology, paleontology, anthropology, geology, and cosmology into a long-term cosmic future. Teilhard de Chardin lays out an unusually long-term cosmic evolutionary trajectory where intelligence plays a significant, if not critical role in the evolution of the universe toward increasing complexity, toward the goal and end point that he called the Omega Point.

In Teilhard, we see ripples of both Herbert Spencer and Henri Bergson, both of who came years before Teilhard, and both who had quite different views about how to interpret Darwinian evolution in light of contemporary human life. Spencer promoted a reductionist vision of human society as being primarily formed by a phrase he coined based on Darwin's *Origin of Species*, i.e., "survival of the fittest" (Spencer 1864). Spencer suggested a kind of universal law of evolution at work in the universe as a whole, for which there was inevitable progress and potentially some kind of end goal, realized at least in part by the evolution of the human mind and human societies (Spencer 1862). Bergson attempted to "soften" the rationalist, reductionist, mechanistic interpretations of Spencer by suggesting that evolution was a kind of spiritual force, noting the development of cooperation and altruism as key features of human evolution (Stevens-Arroyo 2012). If one were to place Teilhard in relation to Spencer and Bergson, it could be said that he lies somewhere between them on a spectrum of rationalism, reductionism, and mechanism on one end (Spencer) and experience, intuition, and spirit on the other (Bergson)—with Teilhard arguably lying closer to Bergson philosophically and metaphysically, but perhaps close to Spencer in envisioning directionality and perhaps even teleology in evolution.

Indeed, Teilhard appears to have been wedded to the idea of an intrinsic directionality in the nature of the world and that matter is intrinsically imbued with spirit and consciousness that unfolds, emerges, and manifests over time in an evolutionary process leading to ever-increasing degrees of "complexification." This kind of directionality is an intrinsic part of the universe as a whole and hence can be seen as a form of "objective directionality" or "intrinsic directionality." It can be seen as a source of what is effectively a general predestined trend or a teleological unfolding—presumably one that may emerge irrespective of whether or not our specific form of human intelligence evolved. So while Teilhard draws heavily from the processes of geological and biological evolution, he goes further than many scientists by imbuing matter and the whole process of evolution with a force or spirit (again, consistent with the views of Bergsen) leading to increasing complexity and consciousness and hence to greater degrees of spiritual realization. Teilhard signals in his writing that consciousness is as important as solid matter and life itself: "Right at its base, the living world is constituted by consciousness clothed in flesh and bone." (Teilhard de Chardin 1955, 151).

Teilhard also notes a critical role for love in evolution. He sees love as a force that unites the world, physically and in other ways (Stevens-Arroyo 2012). This force of love brings elements of the world together to create new forms, often seen as "higher" forms, and helps bring the world into being. For Teilhard de Chardin, this centrally places love in the evolution of the world as something inherent in the world, manifested as a sort of spiritual creative force leading to an evolution of ever-increasing degrees of love.

Frank Tipler, in his 1994 book, *The Physics of Immortality*, builds directly on the ideas of Teilhard de Chardin's *Phenomenon of Man* and offers what appears to be a more mechanistic and much more detailed, perhaps even deterministic, theoretical physical account of cosmic eschatology—claimed by Tipler to be completely consistent with the laws of physics (and notably, somewhat consistent with parts of Christian beliefs). Tipler develops the idea that something like an Omega Point will be realized by the increasing capacities of intelligent beings that culminate in what is essentially infinite information processing power at the end of cosmic evolution. Interestingly, Tipler equates the Omega Point with God in part because the Omega Point possesses some important properties of God as normally conceived, e.g., immortality, benevolence, omnipotence, etc. And this Omega Point God, i.e. this future cosmic society or cosmic mind at the end of universe, can and will, out of purely benevolent love, "resurrect" all beings that ever were as emulations in an infinitely powerful final computational system.

In this sense, Tipler's cosmic mind at the end of cosmic evolution is an eschatological God, a God that comes into being during the course of cosmic evolution. But this God is also consistent with many traditional conceptions of God according to the divine "metrics" of Tipler's assessment, such as benevolence, omnipotence, omniscience, and resurrection.[3] It is a "God of the future," which Tipler takes to be consistent with some interpretations of the Hebrew translation in the Old Testament that suggests a kind of future God who proclaims "I *will be* what I *will be*" (as opposed to "I am what I am"). Tipler claims inevitability for his eschatological God in part because the laws of physics allow it (and perhaps ultimately require it) and also because intelligence will want to survive indefinitely and hence need to effectively consume and "control" the entire universe.

In addition to the motivations of survival, Tipler's eschatological God is motivated by benevolence, by a form of love. This is a love for all beings that have ever existed, who, out of love for them, will be resurrected by this final cosmic mind in the form of extremely high-fidelity emulations to the point where Tipler claims our emulated selves will in fact be us because they will be so quantum mechanically similar to who we are now that there will be no important discernible difference.

2.3 Cosmic Sacredness

Brian Swimme and Thomas Berry also find affinity with many aspects of Bergson's and Teilhard's worldviews. They reframe some of those orientations in light of a more

[3]Tipler's book tends to emphasize consistency with Christianity, in particular, resurrection. But in a personal email correspondence with him, it was unclear why these kinds of common features of many-world religions should be so closely identified primarily with Christianity and not more generally with many other religious traditions.

modern understanding of Earth's biosphere, cosmology, and cultural history, while still maintaining a depth of sacredness and spirit in the cosmos. Swimme and Berry have promoted a deeply celebratory view of the cosmos—not only arguably adopting a kind of pantheism, but reveling in and explicitly celebrating the scientific details of cosmic evolution. They cast cosmic evolution as a grand story that can be a meaningful and perhaps even purposeful narrative that can help guide and transform the human experience, perhaps acting as a form of natural "religion." Swimme sees our knowledge of cosmic evolution as a way to understand our proper place in the world, and as a way to engender humility and respect for our environment.

A revealing detail is that Swimme sees the universe as attempting to be "*felt*," which can be seen in contrast to ideas suggested by some that the cosmos may be attempting to *know* itself via intelligent beings such as ourselves. Swimme does not limit our role to merely awareness or knowing, but extends it, or at least takes a different angle, and suggests a kind of "feeling" is also what the universe is attempting—in part through feeling beings like us.

In a vein similar to Ursula Goodenough's *Sacred Depths of Nature* (1998), Connie Barlow explores how our scientific understanding of nature and the broader cosmos can be viewed and experienced with reverence and awe. In *Evolution Extended* (1994), Barlow seeks guidance from a diverse and well-balanced group of biologists and scientists to explore if and how biological evolution might be indicative of some kind of progress and whether there may be connections to cosmic meaning. While legitimate concerns are presented for reading too much progress and meaning into biological evolution and any connections to cosmic meaning, Barlow still sees room and utility to allow for such explorations. Seeking further, in *Green Space, Green Time* (1997), Barlow provocatively asks "what are humans good for?" and brings us another wonderfully rich array of thoughts and direct quotes from personal conversations she has had with scientists such as Ursula Goodenough and John Maynard Smith as well as provocative philosophers such as Holmes Rolston.

Of particular relevance for this book (particularly the overall relational framework that will be explored later) is Barlow's attention to the idea of a "relational self."[4] She quotes Martin Buber who wrote, "In the beginning is the relation" (Buber 1937, 69). Barlow writes: "In this view of the self, other beings and other aspects of the cosmos are no longer distinctly 'other'; they are relations." (Barlow 1997, 264). A relational self-image can also lead to an expanded sense of self in which the self is seen as much larger than just the traditional sense of self that is often limited to our physical bodies and narrow "local" relationships. A sense of an "embedded self" or "communal self", or "communion" more generally, can help balance between a relational and expanded self-image by seeing all things as both whole and part. Related to these articulations, Barlow suggests a somewhat different but related metaphor for thinking about a broader relational context, namely *conversation*. She writes:

[4]Barlow cites Curtin (1994) who examines self-images and their role in environmentalism and worldviews, one of which is the "relational self."

> Conversation is, after all, a community in process. Conversation binds individuals into communities. Similarly, ecosystems, bioregions, and Gaia emerge and are sustained by a kind of physical and chemical conversation among the living and between the living and non-living. Evolution is an extension of those same conversations through time. Conversation suggests spontaneity, mutual creativity. It has no pre-established destination, yet we can count on something interesting developing. We can give our full attention to the present moment, in effect losing ourselves in conversation, an immense and delightful journey unfolds – effortlessly.

With *conversation* as a kind of cosmic metaphor, and drawing from Swimme and others, Barlow sees humanity as "celebrants of the Universe Story" through conversations and all its myriad forms such as story, art, and song. Barlow quotes Dobb (1995): "It is our immense good fortune and grave responsibility to the sing the songs of the cosmos,"[5] and Barlow quotes Elizabeth Johnson (1997): "We are the cantors of the universe" and sums it up nicely with: "We are the universe celebrating itself. Here the expanded self and joyful expression merge." (Barlow 1997, 271).

Another form of cosmic reverence can be found in "cosmotheology" as articulated by Steven Dick (2000).[6] Like others previously noted, Dick emphasizes the importance of using scientific knowledge to inform and perhaps change theologies in a way that incorporates what science is learning about the cosmos. Dick outlines a number of principles that should guide cosmotheology, ranging from recognizing that humanity is not central to the universe, to being open to radical notions of God that are derived from cosmic evolution, to having a moral dimension that encourages respect and reverence for life. Dick encourages humility in the face of needing to know more about the cosmos, including whether there is life elsewhere, before making specific conclusions about meaning and purpose. But he does conclude the following: "Whether intelligence is rare or abundant, whether life is of a lower order or a higher order than homo sapiens, human destiny is intimately connected with cosmic evolution." (Dick 2000, 205).

2.4 Evolutionary Developmental Universe

The evolutionary developmental nature of evolved or "emergent" systems has been increasingly applied as a model for cosmic evolution. It leverages key distinctions and interactions among stochastic evolution, development, and information processing—similar to biological systems in which basic evolutionary dynamics such as randomness (e.g., mutations) and developmental constraints (e.g., natural selection) shape what emerges. Smart (2009) explores this framework in some detail and leverages suggestions that the universe is a kind of information processing system that is, partly, if not completely, driven by intelligent beings that can give rise to a highly computational system such as a "developmental singularity." Such a singularity might

---

[5]Barlow quotes from Edwin Dobb's 1995 article, "Without Earth there is no heaven," Harper's Magazine, February, pp. 33–41.

[6]Kant appears to have coined the term "cosmotheology" in *Critique of Pure Reason* to capture the idea that a "supreme being" might be inferred by experience of the world. Dick's more contemporary use is different in that it does not require a supreme being.

take the form of a "black hole computing system" (in some ways functionally similar to the Omega Point of the cosmic eschatology noted previously). A black hole can provide a high level of energy density to help produce extreme computational capacities (Lloyd 2000). This kind of black hole computational system, created largely, if not completely by intelligent beings, can then give "artificial" birth to other universes and contribute to a process of cosmological selection similar to that proposed by Lee Smolin (1997) (where constants and laws could possibly vary "randomly"), but with the added component of intelligence, perhaps allowing more control over the nature of those created universes.

Developmental constraints create robust results, often in the form of highly complex systems (such as biological and cultural systems), but do not exceed certain physical constraints that may exist in any particular universe or universal cycle—only relatively small changes are possible from one cycle to the next. Nevertheless, evolution and development, both taken together, forming a kind of computational dynamics, appear to be giving rise to some directionality, perhaps in the form of increasingly complex adaptive information processing systems that will continue to increase in computational capacity.

This apparent, or at least potential directionality may be constrained by a number of factors. Some of those factors can be basic "goals" of complex adaptive systems, captured by the fundamental dynamics of such systems: creativity (evolutionary), adaptive (computational), and developmental (sustaining). Smart suggests: "Consequently, the telos (intrinsic goals/ends/values/drives) of these three processes may be increasingly constraining on a complex adaptive system as a function of their complexity." (Smart 2009, 222). Further, he writes, "We may therefore discover that these three telos act as increasingly powerful constraints on the emergent morality of biological, societal, and technological/post-biological systems." (Smart 2009, 223).

Vidal (2014) extends this thinking a bit further. He suggests an almost one-to-one mapping of evolutionary dynamics to evolutionary ethics and developmental dynamics to developmental ethics, with the distinction remaining similar, if not largely the same. That is, evolutionary values and ethics are essentially based on the stochastic and adaptive short-term practical values that facilitate survival and reproduction and are essentially descriptive, whereas developmental values are more cognitive social long-term drivers of growth and integration and so provide a better foundation for more normative ethics. This leads Vidal to propose an ethic that values "the infinite continuation of the evolutionary process" as an "ultimate good." He writes: "…what matters most are processes that sustain life at large and our universe, rather than the universe conceptualized as a static object."

Even more specifically, what we become aware of and worry about in this ethic "is the importance of the infinite continuation of cosmic reproduction." Vidal continues:

> We are no more attached to our particular universe. …What matters is the recursively fertile infinite production of intelligent universes. In other words, what matters is to make sure that universes continue to evolve and reproduce with intelligence, even beyond our own universe. Thus artificial cosmogenesis becomes a primary way to cognize, not only to decipher to what extent our universe is robust or fine-tuned, but to produce an artificial blueprint for a fertile offspring universe. (Vidal 2014, 291)

As suggested by Vidal, this cosmological reproduction should essentially be infinite in time and hence give rise to a form of "cosmological immortality."

2.5 Participatory Universe

Here we explore what are arguably stronger versions of views touched on previously. The previous views explored roles for intelligent beings to "participate" in cosmic evolution in an intentional manner based on a roughly linear temporal evolution. That is, increasing capabilities and intentional motives of intelligent beings might eventually give rise to the kind of participation we noted above, perhaps ending in an ultimate cosmic intelligence or Omega Point-type cosmic "mind," or perhaps involving participation in a larger multiverse ecology by intentionally creating other universes.

Here we will briefly touch on views based primarily on some provocative interpretations of quantum theory and some experimental results of quantum mechanics. In subsequent sections, we will explore interpretations of quantum theory more closely, but for now we will highlight a broad interpretation of a participatory universe or "co-created" reality rooted in oft-cited interpretation of quantum mechanics that suggests the act of observation plays some role in, if not entirely determines, the results measured—sometimes referred to as the Copenhagen interpretation.

Gardner (2005) suggests a kind of participatory or "co-created" evolutionary model of the universe as a "closed timelike curve" (CTC) which can provide a theoretical explanation for the origin and continued evolution of the cosmos. A closed timeline curve model of the universe can be interpreted as linking past and future states and perhaps more specifically by allowing information, including biological and "cultural" information, to pass through what might be a "big bounce" or "big crunch." Gardner writes: "Because the CTC is curved and timelike and closed and unblemished by a final singularity, each point on the CTC is, to at least a limited degree, both the cause and effect of every other point. Time flows in only one direction in this scenario but because the CTC unites past and future at the Big Crunch threshold, the two temporal states can coevolve." (Gardner 2005, 18). This suggests to Gardner that we can think of our biofriendly universe as a "classic autocatalytic set"[7] that essentially gives rise to its own origin. He elaborates further:

> If our cosmos is indeed a CTC—or if our multiverse is a series of branching CTCs—then the human-dominated past will continue to exert a causal effect on a transhuman future, long after humanity ceases to be the dominant form of intelligence on planet Earth. Likewise, the transhuman future will exert a causal relationship on prior states of the cosmos, including the era of human intellectual supremacy. In the inimitable phrase of John Wheeler, this exotic scenario envisions the whole shebang of past, present, and future not as a traditional sequential history but rather as a 'self-excited circuit' and a 'grand synthesis, pulling itself together all the time as a whole.' (Gardner 2005, 378)

[7] Another invocation of autocatalytic dynamics by Stuart Kauffman will be touched on later regarding the origin of life.

Davies (2009) goes further and adopts John Wheeler's interpretation of quantum theory that is sometimes considered to support the idea of "retrocausation." Davies suggests the possibility that life, mind, and culture are fundamental properties of the universe to the extent that the cultural evolution of observing agents, construed cosmically, is a means by which the universe can become a biofriendly universe—including as *a means by which the laws and constants of the universe are determined*. Davies' suggestion is consistent with the previous eschatological visions of Teilhard de Chardin and Tipler in the sense that those visions suggest that the universe could eventually be consumed and "saturated" by mind. In such a state, the entire universe could be subject to the observer-participancy principle, suggesting that observers participate in creating all of physical reality, including possibly the reality of the universe in the past before there may have been actual physical observers (Wheeler 1990). This general idea is sometimes called the participatory anthropic principle.

In Wheeler's words:

> It from bit. Otherwise put, every 'it'—every particle, every field of force, even the space-time continuum itself—derives its function, its meaning, its very existence entirely—even if in some contexts indirectly—from the apparatus-elicited answers to yes-or-no questions, binary choices, bits. 'It from bit' symbolizes the idea that every item of the physical world has at bottom—a very deep bottom, in most instances—an immaterial source and explanation; that which we call reality arises in the last analysis from the posing of yes–no questions and the registering of equipment-evoked responses; in short, that all things physical are information-theoretic in origin and that this is a participatory universe. (Wheeler 1990, 5)

This principle suggests that physical states of the universe are determined, in part, if not completely, by observers making observations of the physical world, including potential "retroactive" causation into the "past." Davies notes that Wheeler, based on results from the delayed choice experiment (which Wheeler interpreted to suggest that an experimenter affects a previous state of reality), extended this "retroselection" interpretation of the nanoseconds involved in the experiment to billions of years and the entire universe itself. Davies writes:

> The novel feature he (Wheeler) introduced was the possibility that observers today, and in the future, might shape the nature of physical reality that was, in the past, including the far past when no observers existed. It is a radical idea that gives life and mind a creative role, making them an indispensable part of the cosmological story. Yet life and mind are the products of the universe. So there is a logical as well as a temporal loop here. In conventional science one describes a logical sequence: cosmos → life → mind. Wheeler proposed closing this chain into a loop: cosmos → life → mind → cosmos. He expressed it as follows: 'Physics gives rise to observer-participancy; observer-participancy gives rise to information; information gives rise to physics.' In this manner, the universe explains observers, and observers explain the universe. (Davies 2009, p. 394)

So, Davies takes this already controversial idea one very large step further and suggests that if and when the entire universe is effectively consumed by intelligence, the entire universe is influenced by observer-participancy and becomes a kind of universal observer of itself that can then determine not only the physical state of the entire universe, but, as emphasized previously, possibly the laws of physics themselves. Again, it is worth reading Davies' own articulation directly, partly because of the extremely counterintuitive and provocative nature of the claim:

> Building on Wheeler's notion of retroactive observer-participancy, can we explain why the laws of physics are fine-tuned for life? Not in the usual formulation, no. Although quantum mechanics requires the presence of many alternative pasts, every allowed history develops over time in conformity with the same laws of physics. The differences in the histories come about purely from inherent quantum uncertainty, not from any variations in the laws of physics as such. What we would like to explain is why the laws themselves are biofriendly, thus permitting at least some quantum histories containing observers. To achieve this, it is necessary to apply the general principle of linking future to past through quantum observations, but to extend its reach from states of the world to the underlying laws of physics, too. Until now, such an extension would have been meaningless because the laws were regarded as fixed and infinitely precise. But treating the laws as cosmic software, with an inherent flexibility, neatly lends itself to the task. Observations made throughout the entire duration of the universe can contribute to fashioning the form of the laws in the first split second after the Big Bang when they were still significantly malleable. Thus the potential for future life acts like an attractor, drawing the emerging laws towards a biofriendly region of the available parameter space. In this way life, mind, and cosmos form a self-consistent explanatory loop. (Davies 2009, 394)

Davies further writes:

> If the universe were to become saturated by mind (similar to the Final Anthropic Principle (Barrow and Tippler 1994)), then it would fulfill the necessary conditions for Wheeler's participatory principle in which the entire universe would be brought within the scope of observer-participancy. The final state of the universe, infused with mind, would have the power to bring into being the pathways of evolution that lead to that same final state. Thus cultural evolution, on a cosmic scale, would in effect be the lynchpin in the very self-synthesis of the universe, enabling the universe to both create itself and steer itself towards its destiny. (Davies 2009, 395)

Davies seeks to find explanations for the details of the universe that do not require anything outside the universe, but does rely essentially on extreme derivations of the arguably problematic Copenhagen interpretation of quantum mechanics—which will be discussed later in greater detail. Davies' suggestion, however speculative by his own admission, at least conceptually allows for the possibility that the universe can in some sense explain itself via the emergence of sufficiently advanced observers. Such observers will eventually exist at large enough scales to bring the entire universe into observer-participancy, and hence facilitate the emergence of "bio-friendly" cosmic laws that allow intelligent observers to emerge and give rise to what might be called "cosmological observer-participancy" (a more specific and descriptive title than the participatory anthropic principle).

2.6 Science Fiction

As a nod to science fiction, a genre that has provided rich and compelling visions of the future of life and intelligence in the universe, we briefly note one of the earlier modern relevant works on the subject of the far future of life and intelligence in the universe. *Star Maker* was published in 1937 by Olaf Stapledon and is reputed to have been claimed by Arthur C. Clarke to be one of the greatest works of science fiction ever written. It is a different kind of book for sure. It may not be the most accessible science fiction novel, but it is remarkable in scope and emotional depth. Its philosophical underpinnings inspire and provoke the reader to see many questions in a new light, in an emerging cosmic context based on some of the leading science of the day. The book foreshadows a number of concepts that were pursued more thoroughly by others in subsequent years, such as genetic engineering, terraforming, and Dyson spheres. But there is one concept implied by *Star Maker* that is not often traced to Stapledon, namely that of creating many universes, something some physicists have begun to more seriously contemplate (Guth 1994).

One of the many interesting things about Stapledon's fictional treatment is that is not clear whether there is a larger directionality, a broader objective meaning or purpose, or ultimate value to life, intelligence, or the universe. The "creator" itself appears to be confused, or at least unsettled about such things, perhaps reflecting Stapledon's own agnosticism and uncertainty. To the extent that there is purpose or teleology in Stapledon's *Star Maker*, it may be seen in a creator attempting to seek its own satisfaction and meaning through highly diverse creations—including many kinds of beings and many kinds of universes.

The prospect for beings to create universes raises profound questions. Many, if not all of those questions are arguably rooted in what intelligent beings value. If and when it becomes possible for beings to create universes, this will imply the ability to create many alternative physical realities with potentially highly unpredictable outcomes. Values and ethics will presumably inform challenging questions. Should universes be created by intelligent beings, and if so, why? Can we, should we, control those alternative realities? And perhaps most importantly, assuming we can control details of universes that are created, what might those details be? What kind of universes should we create?

2.7 Summary of Chapter 2: Precedent

In this chapter, we explored some worldviews that are relevant to the kinds of worldviews that will be explored later in this book. We touched on contemporary ideas based partly on science and cosmology that offer fairly specific views suggesting notable relevance for cosmology and potentially some form of cosmic significance for life and intelligence. This brief foray into some relevant precedent is intended to act as context and help inform the development of a number of forthcoming cosmological worldviews within a pluralistic philosophical relational framework that leverages fundamental physics, cosmic evolution, and philosophy.

We briefly touched on perspectives such as (1) cosmological eschatology worldviews that emphasize the very far future of cosmic evolution and suggest that intelligence will evolve eventually to the point where it in some sense merges with the universe and possibly creates a kind of eternal "cosmic mind", (2) cosmic sacredness worldviews that emphasize a deep reverence for the cosmos and its evolution as a source of our origins and sustainment, (3) evolutionary developmental views that emphasize the evolutionary developmental nature of the universe and how it may give rise to other universes through the evolution of intelligence, (4) participatory universe views that leverage extreme interpretations of quantum theory to suggest that observation and consciousness give rise to reality, including possibly the laws of physics themselves, and (5) science fiction, with an emphasis on Olaf Stapledon's *Star Maker* which explores the idea of a creator imperfectly creating many universes in an attempt to constantly "improve" on previous creations.

References

Barlow, C. (1994). *Evolution extended: Biological debates on the meaning of life*. Cambridge: MIT Press.

Barlow, C. (1997). *Green space, green time: The way of science*. New York: Springer.

Barrow, J. D., & Tipler, F. J. (1994). *The anthropic cosmological principle*. Oxford: Oxford University Press.

Buber, M. (1937, 1970). *I and Thou* (W. Kauffman, Trans.). New York: Scribner's.

Carroll, S. (2016). *The big picture: On the origins of life, meaning, and universe itself*. New York: Dutton.

Chaisson, E. (2005). *Epic of evolution: Seven ages of the cosmos*. New York: Columbia University Press.

Curtin, D. (1994). Dogen, deep ecology, and the ecological self. *Environmental Ethics, 16,* 195–213.

Davies, P. (2009). Life, mind, and culture as fundamental properties of the universe. In S. J. Dick & M. Lupisella (Eds.), *Cosmos and culture: Cultural evolution in a cosmic context*. Washington, DC: NASA History Series. http://history.nasa.gov/SP-4802.pdf.

De Duve, C. (1995). *Vital dust: Life as a cosmic imperative*. New York: Basic Books.

de Teilhard de Chardin, P. (1955). *The phenomenon of man*. New York: Harper Row.

Deutsch, D. (2011). *The beginning of infinity: Explanations that transform the world*. London: Penguin Books.

Dick, S. J. (2000). Cosmotheology: Theological implications of the new universe. In S. J. Dick (Ed.), *Many worlds: The new universe, extraterrestrial life, and the theological implications* (pp. 191–210). Philadelphia: Templeton Foundation Press.

Dobb, E. (1995). Without Earth there is no heaven. *Harper's Magazine*, February, pp. 33–41.

Dyson, F. J. (1979). Time without end: Physics and biology in an open universe. *Reviews of Modern Physics, 51*(3).

Gardner, J. (2005). Coevolution of the cosmic past and future: The selfish biocosm as a closed timelike curve. *Complexity, 10*(5) (May/June 2005), 14, 17–18.

Goertzel, B. (2010). *A cosmist manifesto*. Humanity+ Press.

Goodenough, U. 1998. *The sacred depths of nature*. Oxford: Oxford University Press.

Guth, A. (1994). Do the laws of physics allow us to create a new universe? In G. Ekspong (Ed.), *The Oskar Klein memorial lectures* (Vol. 2, pp. 71–95). World Scientific Publishing.

Harrison, A. (2013). Russian and American cosmism: Religion, national psyche, and spaceflight. *Astropolitics, 11,* (1–2).

Johnson, E. (1997). Retrieval of the cosmos in theology. *Earthlight*, Spring, 8–9.

Kardashev, N. (1964). Transmission of information by extraterrestrial civilizations. *Soviet Astronomy, 8,* 217.

Lloyd, S. (2000). Ultimate physical limits to computation. *Nature, 406,* 1047–1054.

Lovejoy, A. (1936). *The great chain of being*. Cambridge: Harvard University Press.

Lupisella, M. (2009). Cosmocultural evolution: The coevolution of cosmos and culture and the creation of cosmic value. In S. J. Dick & M. L. Lupisella (Eds.), *Cosmos and culture: Cultural evolution in a cosmic context* (pp. 321–359). NASA SP-2009-4802. http://history.nasa.gov/SP-4802.pdf.

Lupisella, M. (2019). Is the universe enough? Can it suffice as a basis for worldviews? *Journal of Big History, III*(3), 123–140.

Smart, J. (2009). Evo Devo universe? A framework for speculations on cosmic culture. In S. J. Dick & M. Lupisella (Eds.), *Cosmos and culture: Cultural evolution in a cosmic context*. Washington, DC: NASA History Series.

Smolin, L. (1997). *The life of the cosmos*. New York: Oxford University Press.

Spencer, H. (1862, 1898). *First principles of a new system of philosophy*. New York: D. Appleton and Company.

Spencer, H. (1864). *Principles of biology* (1867; 2nd ed., 1898–99). London: Williams and Norgate.

Stapledon, O. (1937). *Star maker*.

Stevens-Arroyo, A. M. (2012). Fire and force: Civilization as noosphere in the works of Teilhard de Chardin. *Comparative Civilizations Review, 66*, Spring.

Swimme, B., & Berry, T. (1992). *The universe story: From the primordial flaring forth to the Ecozoic Era: A celebration of the unfolding of the cosmos*. Harper.

Tipler, F. (1994). *Physics of immortality*. New York: Doubleday.

Vidal, C. (2014). *The beginning and the end: The meaning of life in a cosmological perspective*. Berlin: Springer.

Wheeler, J. A. (1990). Information, physics, quantum: The search for links. In W. Zurek (Ed.), *Complexity, entropy, and the physics of information*. Redwood City: Addison-Wesley.

Whitehead, A. N. (1929). *Process and reality: An essay in cosmology*. New York: Macmillan. Edition 1978 by D. R. Griffin & D. W. Sherbourne, New York: Macmillan.

Chapter 3
From Replication to Aspiration

We are all in the gutter, but some of us are looking at the stars.
—*Oscar Wilde*

3.1 Background and Overview

In much of what was briefly explored in the previous chapter, we see claims for roles of human agency—or intelligent agency more generally—in cosmic evolution. The desires and values of intelligent agents, including possibly machine intelligence or artificial intelligence, can critically shape and drive choices of potentially very powerful beings and hence can influence a much larger whole, including possibly the entire universe (or universes). Here we will briefly touch on some sources of values, and value theory more generally, to help inform our subsequent exploration of a variety of scientific and philosophical perspectives that can help then inform cosmological theories of value.

3.2 Biology and Intelligence

Biological systems can arguably be seen primarily as blindly voracious replication machines (Dawkins 1989). While biological systems have evolved to produce controlled, cooperative, and even "altruistic" behavior, it is nevertheless fundamental to natural selection to stumble upon means by which genetic propagation is enhanced. Balances among harmful actions that can compromise genetic propagation into the future are struck by Darwinian evolution in many ways, but the process of natural selection is crude and unreliable, and most importantly, it is almost entirely in service of achieving genetic replication into the next generation. This suggests that highly capable life-forms can arise, both unintelligent and intelligent, that have very little, if

M. Lupisella, *Cosmological Theories of Value*, Space and Society,
https://doi.org/10.1007/978-3-030-25339-4_3

any, awareness—and yet can be remarkably capable in terms of survival and reproduction fitness. This further points to the possibility that there may be highly capable and intelligent beings throughout the universe that could exhibit blindly extreme behaviors.

Intelligence seems to have evolved primarily, if not completely, as a mechanism to enhance genetic fitness, perhaps as a result of complex social living or other conditions that require significant information processing and highly adaptive brains (Sternberg and Kaufman 2002). Intelligence could be nothing more than a high degree of competence without understanding. Dan Dennett has suggested that we educate children so they can understand in order to be competent (Dennett 2012). I would like to suggest that the reverse is also true from a Darwinian perspective. That is, because of a certain level of competence (e.g., "general intelligence") already programmed into us by evolution to help us survive and reproduce, certain species also have a basic capacity to "understand." And how we humans understand has much to do with culture.

3.3 Culture

Most culture is arguably shaped by natural selection (Barkow et al. 1992). Evolutionary psychology, with its important connections to group selection and social psychology, has much to say about human psychological predispositions (including predispositions to seek broader contexts for meaning and purpose!—something explored further in subsequent chapters). Based on the premises of evolutionary psychology and group selection, like intelligence, culture has arguably evolved primarily as an adaptive mechanism to enhance genetic fitness of individuals and groups without necessarily requiring a high degree of awareness. However, modern culture (say within the last 10,000 years since around the beginning of effective agriculture)[1] has arguably produced a remarkable level of awareness.

This awareness of self and other minds seems to almost certainly imply some form of internal modeling of one's self in a larger context and/or rough "theories of mind" to allow for a more effective awareness of others and hence more effective interactions with others—leading ultimately more stable and effective groups. This more immediate practical awareness might lend itself to an increasing deepening of self-awareness, for example leading to an awareness of one's own death and the death of others—perhaps then leading to an increasing awareness of alternative possible future states for one's self and one's group. The combination of both immediate and longer-term future-oriented awareness, along with the ability to imagine, contemplate, and communicate with others about those future states, has increased our

[1] It is hard to say when "modern" human culture started to develop, but we might think about it starting around the time when agriculture began to produce more stationary groups, hence allowing for more free time for some members of those groups, more complex division of labor, more sophisticated cooperation, and more detailed means of accounting for and controlling resources and behavior.

awareness at many levels (including an awareness of our awareness!) and may be a big part of what constitutes "consciousness."

The ability to simulate and explore future states with others has arguably led to an explicit awareness of the utility of proactively seeking and creating favorable future states. This explicit awareness would presumably benefit greatly from the myriad cultural constructs and strategies that help realize those states. This would include systems of intellectual thought that help optimize the pursuit of favorable future states for both individuals and groups, which can be thought of as philosophical systems of thought, with theories of value being a key subset. Culture, then, can be seen as a kind of bridge between intelligence and the *normative aspiration* of sufficiently aware beings.

3.4 Value Theory and Normative Aspiration

Human culture has led to broad and deep levels of philosophical inquiry, a major area of which is value theory. In its simplest form, value theory carefully and rationally explores what is valuable and why, and includes, for example, ethics and aesthetics. "Normative aspiration" can be thought of as a proactive cognitive pursuit of what is valuable or ideal based on consideration of alternative future states, from which something is cognitively chosen as a "norm" to aspire to. Like intelligence and culture, normative aspiration is also highly influenced by Darwinian evolution. As David Hume suggested in 1738 in *A Treatise of Human Nature*, reason is likely a slave to the passions, suggesting that our emotional predispositions (our passions) are really what drives our reasoning about what is valuable—which is presumably focused on what is valuable based on feelings that were selected for by evolution to enhance fitness. Hume did not know about evolution by natural selection and so did not have an appreciation for the specific natural mechanism that could produce such strong passions, but Hume's observation points us to an important consideration.

In contemporary human life, normative aspiration increasingly appears to be more than just aspiring to what our passion-driven reasoning leads us to. Philosophers and others from many disciplines have conducted numerous forays into the realm of "theories of value"—many of which involve notable attempts to be rational, fair, and objective in assessing what is valuable and how we might realize it (e.g., Kant's "categorical imperative" (Kant 1785) and John Rawls' "veil of ignorance" (Rawls 1971)—despite legitimate concerns of "rational delusion" (Haidt 2012). There is no shortage of ideas in value theory, many often completely contradictory, sometimes creating the misguided impression that values are so subjective and arbitrary that they are not amenable to rational analysis (Putnam 2002).

But how human beings value has much, if not everything, to do with individual internal psychological states. Scientists continue to learn much about those states and their relation to evolution and the external world, to the point where there is increasing hope for a more empirical scientific grounding of moral psychology, ethics, and value theory in general (Shermer 2004; Harris 2010). Related to this more empirical approach to normative aspiration is the interplay between what are often distinguished as "facts" and "values." Indeed, one way of thinking about normative aspiration is that it can help lead to wisdom and perhaps ultimately the creation of a deep reality that reflects that wisdom—perhaps a simple "formula" like: Facts + Normative Aspiration $\Longrightarrow$ Wisdom

So how do we cope with this very large, if not infinite, landscape of normative aspiration for many different kinds of beings? Can we, how might we, rationally navigate it? How might we view fact–value interactions? There are several areas of science or "facts" that can play useful roles in informing our thinking about values. Here we will briefly touch on normative aspiration in the scientific contexts of natural selection and cosmic evolution.

3.4.1 Fact–Value Interplay

Because we will be using our knowledge and perceptions of what appear to be facts about the world to inform our philosophical discussion of values and normative aspiration, it will be fruitful to spend some time on what might be thought of as the "fact–value interplay." There is a long-standing distinction in value theory between facts and values. Related to that distinction is an observation emphasized by Hume in 1738 suggesting that values (or "oughts") cannot be necessarily derived by something so different as facts (or the way the world "is"). The "naturalistic fallacy" also cautions against making conclusions about what is valuable or good based on natural properties (Moore 1903).

It is prudent to make note of ideas like the naturalistic fallacy because this book rests partly on the idea that our knowledge about the universe might usefully inform our perceptions of value and our normative aspirations. However, while aspirations can be very much informed by what we learn about the world, they do not necessarily need to be constrained by perceptions of reality or "facts." It would not be unreasonable to constrain conclusions about values and ethics to what we think are real facts about the world—it is an understandable choice for a variety of reasons, and this will be discussed further in later sections. However, it is also not unreasonable to use perceptions of reality to significantly inform our explorations of values and ethics while not necessarily being constrained solely by such facts—as the naturalistic fallacy and fact–value distinction suggest.

There are many ways to think about how facts and values can relate to each other. Two of those ways could be categorized as (1) realism considerations, i.e., whether values are "real" in the sense that they are "objective," and (2) influence considerations, i.e., if and how facts should influence values and vice versa. Realism is the philosophical tradition that asserts that what is real about the world is objectively real in the sense that is ultimately independent of human beings. Science often lays claims to realism, whereas values and morals are often thought to be entirely a product of the human mind, and in that sense, "subjective." Moral realism suggests that there are indeed moral "facts" about the world that are essentially discovered by human beings, similar to scientific physical facts (Devitt 1997). While the question of realism is relevant to this discussion, the emphasis here will instead be on how facts and values might interrelate, and more importantly, how each can be influenced by the other.

In thinking about how facts and values relate to each other, we might first emphasize that they are just different enough to be worth our careful consideration for how they may influence each other. Historically, the distinction seems to have been both overemphasized and oversimplified, suggesting at times that (a) facts do not or should not have much to do with values—derived partly, and perhaps incorrectly, from David Hume's original "is–ought," "fact–value" articulation noted above, (b) that values are not in any way facts, as suggested by some versions of moral relativism, or (c) that facts can and perhaps should determine values, as suggested by more contemporary scientific treatments (e.g., Harris 2010).

The position adopted here is essentially twofold: (1) Facts certainly should significantly inform, but not completely determine values, and (2) values can increasingly influence and determine external physical reality (e.g., "facts") as a species becomes more capable. This implies that while science may not be enough to completely or properly determine values, it is very important as a source of information to help us assess values and means for realizing those values. Science, and the scientific method more specifically, is particularly relevant for the implementation of the goals we value. But the ends themselves, our ultimate values, have a normative component, an aspirational element that may be determined by science (or other "facts") or not. It is essentially a choice—a choice that should be based on careful individual and collective reasoning, a choice that should be made in light of what we have learned about human behavior and our wider world. But we should acknowledge that facts and science may underdetermine values and ethics. Facts primarily occupy an empirical space. Values can occupy a logical possibility space. Facts are rooted in the present. Values can be much more about the future. Facts may not be very flexible. Values are highly malleable.

So it seems there are at least three ways we can think about how facts and values relate and influence each other: (1) Facts do and should *determine* values, (2) facts do and should *inform* values, and (3) facts have little, if anything, to do with values. The latter might also be expressed as facts being only marginally relevant to values, if at all, or alternatively, the way the world "is" does not have much relevance to how it "ought" to be. It seems reasonable to emphasize the moderate "middle way" that facts inform values but do not necessarily constrain or completely determine values. Perhaps the more interesting and important question is *what kind* of influence facts should have on values—which often depends on the specific issue being considered.

3.4.2 Evolution and the Fact–Value Interplay

To explore the fact–value distinction in more concrete detail, with an example that will be important in subsequent chapters of this book, we can start with evolution by natural selection. The scientific basis of evolutionary selection forces (including group selection as an important force for shaping group and social living) and the resulting architectures of brains are facts that appear to heavily influence many, if not most (or perhaps even all?) basic values for contemporary humans—including moral predispositions (Wright 1994; Shermer 2004; Curry 2006; Harris 2010; Haidt 2012). At a high level, it is a fairly straightforward story in the sense that behavioral traits that result in the propagation of an individual's genes into the next generation are the behavioral traits that will tend to endure over time. Such behaviors are manifestations of self-interested motivations, or what can be thought of as "evolutionary values," that are essentially encoded in our genes and manifested in emotions, thoughts, and behaviors that increase the probability that genes will be replicated into the next generation.

Self-interest then is presumably a critical foundation for behavioral motivations, and self-interested behavioral motivations are often driven by evolutionary "values" and related derivatives such as cultural values, which arguably serve evolutionary purposes as well. The behavioral motivational values that promote genetic self-interest, which includes altruism toward relatives (who share genes), and that promote altruistic acts toward others that are considered to be in-group members that can help advance survival and reproduction fitness, will almost certainly be values that endure—including as values that will be promoted in cultural norms. Individual behavioral traits and their associated motivational values that increase the probability of a group persisting can also strengthen and endure over time as groups compete against the environment and each other.

So in this sense, values can be seen as a kind of evolutionary biological fact—or they can at least be "derived" from biological "facts" and science more generally as we learn more about biological evolution and cognitive architectures. These values may be considered facts in the sense that values are representations and manifestations of naturally developed brain states—neuronal patterns in the brains of animals designed to increase genetic fitness. This is one of the way of thinking about how many values are effectively facts that are not just limited to human minds, but that represent facts of biology more generally.

Perhaps most interestingly, evolution by natural selection can be thought of as a logical physical necessity of sorts—a form of "filtered replication" whereby replicative entities are filtered out depending on their ability to successfully replicate.[2] If this is so, and if biological values reflect the behaviors of replicating animals, then such evolutionary or "replicative" values are arguably a basic reality of the world. Wherever there is replication that can endure, vary, and be filtered out and/or promoted based on certain conditions, crude forms of "values" will emerge that motivate behavior for replicative ends, including the behavior of any replicator in principle

[2] Indeed, the phrase "natural selection" may give the phenomenon too much credit—as though an "external" thing called nature is intentionally "selecting" for something—when really it is the much simpler matter that whatever is successful at replicating continues to replicate. Being so close to a tautology without actually being a tautology is arguably a measure of its powerful simplicity. Natural selection is arguably one of the most powerful ideas in human history.

(Blackmore 2009).[3] So in this sense, many values are essentially biological facts, and many values emerge (culturally or otherwise) as a result of biological motivations.

So, many or most values are critically, if not completely, derived from, or at least directly influenced by, scientific biological facts. And this can happen intentionally as beings become more aware of evolution by natural selection. That is, beings aware of their evolutionary origins can choose to stay constrained by biological values (e.g., selfishness and/or limited altruism) because they see it as a fundamental and immovable fact of the world. Related, as shown in Fig. 3.1, beings may choose to stay constrained by certain facts such as biological selfishness because it may ultimately be too difficult to resist or overcome in the ways that might be needed. Indeed, selfishness may ultimately be a kind of "selfishness trap" that is extremely difficult, if not impossible, to evolve beyond (Lupisella 2001).[4]

The previous example using evolution by natural selection suggests that values can be tightly bound to facts, so that facts can, in an important sense, determine values. However, as noted prior, what is most important for our considerations here is to explore the option that while facts certainly can and perhaps should inform values, they do not necessarily need to constrain values. This relates to Hume's point. Strictly speaking, Hume arguably made what was a relatively narrow point of logic, namely that value-based conclusions, do not *necessarily* logically follow from factual premises because factual premises (which are supposed to be "objective") are fundamentally different from value premises (which are largely subjective). It is arguable that what Hume observed was that many writers had too easily slipped into deriving *necessary* value conclusions based on factual premises that are different enough to prevent necessary value-laden conclusions based solely on factual premises. It is

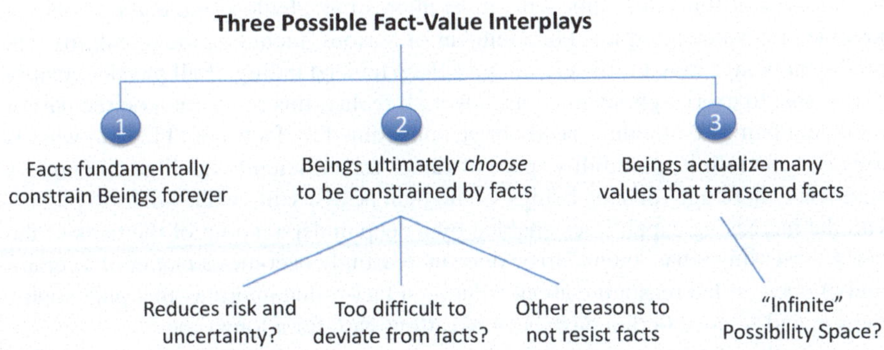

Fig. 3.1 Fact–value interplays

[3]Susan Blackmore, in a Chapter titled "Dangerous Memes; or, What the Pandorans let loose", suggests the emergence of a replicator she calls "temes'—i.e., technology that copies, varies, and "selects" information (e.g., computers, the Internet, etc.)

[4]In *Co-Creating a Public Philosophy for Future Generations*, editors Tae-Chang Kim and Jim Dator note in the introduction how deep the challenges are in developing sufficiently responsive philosophies to future generations.

debatable whether Hume meant to suggest that values absolutely could not be based on facts. The position taken here is that values can indeed be informed by, and based on facts if sufficient science, careful reasoning, and informed choices of intelligent beings lead them there.

3.4.3 Fact–Value Working Assumptions

As noted previously, a working assumption for this book is that value judgments can indeed follow reasonably from factual premises if the reasoning and facts justify it. If the reasoning is sufficient to persuade enough individuals with careful consideration and exploration of many options, that may be good enough. But this is in some sense a methodological, or even epistemological, choice. Such a choice might be seen to have sufficient historical evidence to justify it, but others may not see sufficient evidence, in which case, it is ultimately a choice that depends on a number of factors as Fig. 3.1 indicates. It may be that there are too many unknowns in deviating from facts (as an intelligent species may understand them), or that it is just too difficult to deviate from facts in order to realize certain values as the first two branches in Fig. 3.1 suggest. It may also be that it is undesirable to deviate too much from facts for other reasons—such as ideology or other considerations tied to philosophy and worldviews.

Another working assumption for this book is what is shown as the third branch of Fig. 3.1, namely that beings may ultimately believe and be able to actualize the possibility that values are not, or at least *should not*, be constrained by facts. For the purposes of this book, this is the most interesting, challenging, and compelling possible fact–value interplay. For a number of reasons, including the possibility that intelligent beings may ultimately obtain a deep understanding of all physical reality and be able to exert significant control over all reality, this realization on the part of intelligent beings that values need not be constrained by facts could lead to what is effectively an infinite possibility space of values for sufficiently intelligent beings. A significant challenge for such beings would then be assessing what values to pursue, with the increasing capabilities enabled by a deep understanding of the facts of the world. Assessing what "ought" to be done increasingly becomes a matter of a certain kind of open-ended reasoning about values—a fact–value interplay that philosophy, science, and science fiction have been wrestling with for a long time.[5]

To take this challenge one step further, we can explore briefly how the fact–value interplay may evolve over time (Lupisella 2015). Despite many important counterarguments, we can consider the possibility that, in general, any present state of

[5]Kim Stanley Robinson made this clearer to me many years ago when he said the science in science fiction often represents facts, while the fictional part usually represents values, hence "science fiction" often explores various forms of fact-value interplays. That's when I learned a key reason why I like science fiction!

human values is largely constrained by facts about the world—not just the fundamental facts of physics, but also, and perhaps more interestingly, the facts about the construction of human brains by various forms of biological and cultural evolution. Many open-ended artistic pursuits seem to better approach the slowly increasing distance between values and facts, at least as representations of deviations that can exist in many human minds as potential unknown present and future states—and possibly even as visions that could be physically realized in the future (hence the importance of a genre of literature such as science fiction). This stretching of values beyond facts is captured in much art and literature, sometimes explicitly (as is often the case in science fiction), and often implicitly.

So as the capabilities of intelligent beings increase, such agents may be less constrained by facts and possibly much less interested in facts. Values may come to completely dominate the lives of sufficiently intelligent and capable beings, both culturally and physically. Indeed, it appears this is happening to our species. We are increasingly able to create a world that deeply satisfies our desires, our values—on practically all levels. This raises important questions regarding how humanity might navigate an increasingly complex and diverse (possibly infinite?) possibility space of values, particularly in light of our increasing capabilities. Subsequent chapters will explore a few worldviews that may be brought to bear on this kind of challenge.

3.4.4 Instrumental and Intrinsic Value

Another notable challenge in value theory is whether and how to "locate" value(s) and their source(s)[6], and from that, to at least partially derive some rationally justifiable value(s). Viewed broadly, two themes feature strongly in value theory and ethics: deontology and teleology. A simplified but useful way to distinguish between the two is that deontology generally emphasizes the value of acts, whereas teleology emphasizes the end or consequences of acts (Nozick 1981; Pojman 1995). Notably, both approaches appear to ultimately involve some appeal to basic forms of value. Teleological views, while focusing on consequences, must ultimately give some account of why certain consequences or ends are preferred or valued over others. Some form of fundamental value, perhaps something akin to intrinsic value, can be helpful regardless of which approach is adopted.

The idea of truly *intrinsic* value—value that is independent of a valuing agent—seems difficult to justify (Smith 2009) despite its obvious usefulness for human morality and ethics (particularly metaethics). What does seem reasonable to defend is that value can often be understood at least in a relational sense where value is often clearly realized in situations where something is of value to something else. This kind of relational view of value can be one basis for what is often thought of as

[6]Abraham Odel notes that when fact is interpreted to "embrace the total network of truth, from which it passes readily to encompassing the whole of reality," the fact–value problem is "almost converted into 'Where in the world are we to locate value?'" (Odel 1980, 4).

"instrumental" value. We know things are valuable to human ends as a function of how things are related to human interests or biological interests more generally, and as such, those things at least have instrumental value for us. Intrinsic value is a harder concept to justify, but we will explore some potential justifications in subsequent chapters, based partly on considerations of cosmic evolution—particularly Chap. 8 on the connection-action principle.

3.4.5 The Psychology of Philosophy

> I believe that psychology sits at the center of intellectual life. In one direction, it looks to the biological sciences, to neuroscience, to genetics, to evolution. But in the other, it looks to the social sciences and the humanities.[7]
>
> —Steven Pinker

Much of what is bound up in considerations of facts versus values, the naturalistic fallacy and intrinsic value, and arguably philosophy in general, are factors of human psychology (Bartlett 1989). It may be informative to very briefly turn the lens of psychology—initially with an emphasis on evolutionary psychology—on philosophical explorations.[8] For example, human tendencies to oversimplify and find patterns, meaning, and purpose, including those that are aesthetically appealing, are infamously problematic—a subset of which is sometimes referred to as "promiscuous teleology" or even tenacious teleology (Kelemen et al. 2013). Such tendencies may have evolutionarily advantageous for many reasons related to fitness—perhaps most related to social status and group selection. There are arguably many other evolutionary psychological factors to consider that shape our tendencies to seek trends and patterns, and meaning and purpose. For example, the need to find definitive answers, and do so quickly, can help facilitate "closure" (Kruglanski and Webster 1996) and provide guidance for individuals and groups to help facilitate stability and action. There can also be sources of bias from life history or gender influences (Jacobson 2004).

So do rational appeals work? Or are passions ultimately indeed slave to reason as Hume suggested? Despite our attempts at fairness and justice, it appears most human behavior is still unfair and unjust on many levels—largely dominated by behavior stemming from the genetic evolution of self-interest. Can we then hope to

---

[7]Colleen Walsh interview with Steven Pinker, "What could be more interesting than how the mind works?" Harvard Gazette, May 6, 2014. https://news.harvard.edu/gazette/story/2014/05/what-could-be-more-interesting-than-how-the-mind-works/.

[8]Eric Schwitzgebel has periodically touched on the "psychology of philosophy" in his Splintered Mind blog: http://schwitzsplintersunderblog.blogspot.com/2008/06/psychology-of-philosophy.html?view=magazine. There are journals that explore the intersections between philosophy and psychology such as the *Journal of Philosophical Psychology* and the *Journal of Theoretical and Philosophical Psychology*. There is even a Society for Philosophy and Psychology (http://www.socphilpsych.org/).

make good attempts at, or good use of, rational worldviews, especially those that make "foundational" claims and appeal to something as big and unknown as the universe?[9] Perhaps more unsettling, can we have confidence such worldviews are indeed rational and fair versus ultimately driven by "cleverly" deceptive emotionally laden self-interest in the service of genetic replication? Are our memes blindly doing the bidding of genes? Rational philosophy presumably attempts to avoid reasoning being controlled by emotions, but if it is ultimately unavoidable in some deep sense, it may ultimately be futile to try. It may be that most of us, if not all of us, cannot ever really transcend the narrow forces of biological evolution and the tricks our minds play on us—tricks that do the bidding of our blind, but nevertheless very effective, selfish genes. Despite the many imaginable ways out of this, including perhaps ultimately, artificial intelligence, we may be in a kind of trap that we can never truly escape.

Indeed, many of our predispositions produced by biological evolution presumably operate subconsciously and without careful reasoning or commitments to truth-value (Wilson 2002). Such tendencies have many advantages and disadvantages of course, including how easily we can mislead ourselves for what could be evolutionarily advantageous reasons—again, something that most brands of philosophy are supposed to work hard to avoid. The pursuit of grand philosophical "theories of everything," particularly those attempting to link disparate areas of science with values, should be particularly sensitive regarding potentially problematic psychological motivations. For example, predispositions for seeking value and weighing degrees of value in order to make "better" decisions can have complex selfish and group selection causes, including improving social status and group fitness—even at great cost.

Many worldviews, particularly those that emphasize progress, or that are deductively purpose-driven (including those explored in this book), can be understandably criticized for claiming to find, or at least appealing to something for which there is insufficient evidence. Perhaps worse, such views can be legitimately criticized for a certain kind of wishful thinking or self-deceit—something Peter Medawar accused Teihard de Chardin of (Medawar 1961). So bearing these cautions in mind, let us nevertheless carefully tread into the unsettled waters of meaning-making.

---

[9] We should make brief mention of philosophical schools of thought that question whether "foundations" or "fundamentals" in philosophy can be, or need to be, established at all—either epistemoloical or "metaphysical". Questions regarding "foundationalism" and "fundamentality" should be taken seriously and have been explored in a variety of forms (Rockmore and Singer 1992, Bliss and Priest 2018). Indeed, as cautioned in this brief section on the "psychology of philosophy," the pursuit of a basic fundamental all-encompassing epistemology and "ground" of reality could reflect psychological needs more than it may reflect what is real or what can be known. Gergen (2014) proposes "process without foundation" and advocates a form of process metaphysics—a hint of things to come in subsequent chapters.

3.5 Summary of Chapter 3: From Replication to Aspiration

In this chapter, we briefly touched on some key elements of basic science such as biological evolution leading to intelligence and cultural evolution and how those developments can ultimately help lead to the normative aspiration of cultural beings like ourselves. We explored some fundamental areas of value theory such as the fact–value interplay and how the values of cultural agents may ultimately become a major force in cosmic evolution. We also touched on the distinction between instrumental value and intrinsic value, noting the relational nature of instrumental value and the difficulties of justifying intrinsic value—which will nevertheless feature strongly in the chapter on the connection-action principle. Finally, we broached the subject of the psychology of philosophy, sensitizing ourselves to how psychological predispositions can motivate and possibly mislead us in the pursuit of finding trends and patterns, meaning and purpose, and "grand theories of everything."

References

Barkow, J. H., Cosmides, L., & Tooby, J. (1992). *The adapted mind: Evolutionary psychology and the generation of culture*. Oxford: Oxford University Press.

Bartlett, S. J. (1989). Psychological underpinnings of philosophy. *Metaphilosophy, 20*(3 & 4), 295–305.

Blackmore, S. (2009). Dangerous memes; or, what the Pandorans let loose. In: S. J. Dick, & M. Lupisella (Eds.), *Cosmos and culture: Cultural evolution in a cosmic context.*

Bliss, R. & Priest, G. (Eds.) (2018). *Reality and its Structure: Essays in Fundamentality*. New York: Oxford University Press.

Curry, O. (2006). Who's afraid of the naturalistic fallacy? *Evol Psychol, 4,* 234–247.

Dawkins, R. (1989). *The Selfish Gene*. Oxford: Oxford University Press.

Dennett, D. (2012). *A perfect and beautiful machine': What Darwin's theory of evolution reveals about artificial intelligence*. The Atlantic, June 22, 2012. https://www.theatlantic.com/technology/archive/2012/06/-a-perfect-and-beautiful-machine-what-darwins-theory-of-evolution-reveals-about-artificial-intelligence/258829/.

Devitt, Michael. (1997). *Realism and truth*. Princeton: Princeton University Press.

Gergen, K. J. (2014). Process without foundations. *Journal of Theoretical and Philosophical Psychology, 34*(3), 195–199. https://doi.org/10.1037/a0035424.

Haidt, j. (2012). *The righteous mind*. New York: Random.

Harris, Sam. (2010). *The moral landscape: How science can determine human values*. New York: Free Press.

Hume, D. (1738). *A treatise of human nature*.

Jacobson, A. J. (2004). The psychology of philosophy: Interpreting Locke and Hume. In L. Alanen & C. Witt (Eds.), *Feminist reflections on the history of philosophy. The new syntheses historical library (texts and studies in the history of philosophy)* (Vol. 55). Dordrecht: Springer.

Kant, I. (1785). *Foundations of the metaphysics of morals*.

Kelemen, D., Rottman, J., & Seston, R. (2013). Professional physical scientists display tenacious teleological tendencies: Purpose-based reasoning as a cognitive default. *Journal of Experimental Psychology: General, 142*(4), 1074–1083.

Kim, T.-C., & Dator, J. (1999). Future generations: They are our conscience. In K. Tae-Chang & J. Dator (Eds.), *Co-creating a public philosophy for future generations* (pp. 1–14). New York: Praeger.

Kruglanski, A. W., & Webster, D. M. (1996). Motivated closing of the mind: "Seizing" and "freezing". *Psychological Review, 103*(2), 263–283. doi.apa.org/journals/rev.

Lupisella, M. (2001). *Participant statement in Humanity 3000 Seminar No. 3 Proceedings*, 37. Bellevue, WA: Foundation for the future. See also pages 251 and 330 for additional references to "selfishness trap." Accessed December 28, 2012. http://www.futurefoundation.org/documents/hum_pro_sem3.pdf.

Lupisella, M. (2015). Life, intelligence, and the pursuit of value in cosmic evolution. In S. J. Dick (Ed.), *The impact of discovering life beyond earth*. Cambridge: Cambridge University Press.

Medawar, P. (1961). Reviewed work of the phenomenon of man by Pierre Teilhard de Chardin. *Mind, 70*(277), 99–106.

Moore, G. E. (1903). *Principia ethica*. Cambridge: Cambridge University Press.

Nozick, R. (1981). *Philosophical explanations*. Cambridge: Harvard University Press.

Odel, A. (1980). *Exploring fact and value*. London: Transaction Books.

Pojman, L. P. (1995). *Ethical theory: Classical and contemporary readings*. New York: The Wadsworth Publishing Company.

Putnam, Hilary. (2002). *The collapse of the fact/value dichotomy and other essays*. Cambridge: Harvard University Press.

Rawls, J. (1971). *A theory of justice*. Cambridge: Harvard University Press.

Rockmore, T., & Singer, B. J. (1992). *Antifoundationalism Old and New*. Philadelphia: Temple University Press.

Shermer, Michael. (2004). *The science of good and evil*. New York: Times Books/Henry Holt & Company.

Smith, K. C. (2009). The trouble with intrinsic value: A primer for astrobiology. In C. Bertka (Ed.), *Exploring the origin, extent, and future of life: Philosophical, ethical and theological perspectives* (pp. 261–280). Cambridge: Cambridge University Press.

Sternberg, R. J., & Kaufman, J. (2002). *The evolution of intelligence*. New York: Psychology Press.

Wilson, D. S. (2002). *Darwin's cathedral: Evolution, religion, and the nature of society*. Chicago: The University of Chicago Press.

Wright, R. (1994). *The moral animal*. New York: Vintage Books.

Chapter 4
Scientific Minimalism

4.1 Background and Overview

The provocative speculation of some of the precedent previously explored in Chap. 2 can be balanced by what might be called "scientific minimalism." As shown in Fig. 1.1, the science explored here can help inform and lead into cosmological worldviews that will leverage much contemporary science. Scientific minimalism can be characterized for our purposes by a few basic interrelated methodological ideas and interpretations such as: epistemological prudence, functionalism, and increasing verisimilitude. We will briefly touch on some high-level value theory implications of scientific minimalism and then follow with a fairly in-depth exploration of some key contemporary science such as quantum theory and its potential implications for the more extreme applications to cosmic origins and cosmic evolution, followed by a brief foray into the psychology of science.

4.2 Epistemological Prudence

Epistemological prudence can be thought of as having two parts. The first is often referred to as "Occam's razor" which suggests that all things being equal, the fewer assumptions a theory holds, the better the theory is likely to be when compared to others. So in general, for competing theories, the theory that makes the least assumptions is preferable. This is a kind of parsimony argument, perhaps even an aesthetic argument, that values simplicity over complexity (and relates to Daniel Dennett description of functionalism in the next section). This view reflects a desire to keep theories as simple as possible, but no simpler.

M. Lupisella, *Cosmological Theories of Value*, Space and Society,
https://doi.org/10.1007/978-3-030-25339-4_4

A second element of epistemological prudence is reserving judgment about ideas and theories in the absence of sufficient evidence—with an emphasis on independently verifiable empirical evidence.[1] This suggests refraining from conclusions about the "essence" of the world, particularly strong physical, ontological, or metaphysical claims, since there is presumably often, if not always, insufficient evidence to support such claims. While much of logical positivism or "logical empiricism" is thought to have been undermined since its articulation in the early twentieth century (Hanfling 2003 provides a history), it can still be interpreted as being related to scientific minimalism in the sense that it stresses the importance of empirical and logical verifiability of ideas and theories as a way to discern their worthiness for attention as candidates for useful knowledge (Ayer 1936; Friedman 1999).[2]

4.3 Functionalism

Related to epistemological prudence and increasing verisimilitude that follows, functionalism is arguably a foundational methodological principle that is often not explicitly explored—perhaps in part because it is arguably so operationally fundamental to the scientific enterprise. Like epistemological prudence, functionalism can also be seen to have two aspects. First, what matters in science is what is functionally, or operationally, useful in the theoretical and empirical practices of science. These theoretical or practical constructs need only to be able to provide some useful function in a theory or experiment.

Second, and related to the operational utility noted above as well as the epistemological prudence noted previously, functionalism suggests that a system's behavior is often much more important than what something is fundamentally constituted by, i.e., what it is made of. It is what reality *does* that matters—how it behaves. It is reality's dynamics, how it *acts,* that is most important for most fundamental scientific theories and empirical pursuits. What is most valuable is what is operational by many investigators working independently—suggesting that something about the real world is being learned when things appear to operate roughly the same to independent investigators. Scientific minimalism is more about operational and functional knowledge than it is about fundamental truth. Functionalism suggests that what matters most are interactions at all levels, at fundamental levels of reality, including the roles played by scientists and observers more generally (a topic which will be explored further in this chapter). Daniel Dennett captures functionalism in this way:

> Functionalism is the idea that handsome is as handsome does, that matter matters only because of what matter can do. Functionalism in this broadest sense is so ubiquitous in science

[1] For example, Stephen Hawking has wondered about the prospects for theoretical physics given the challenges of obtaining sufficient evidence (Hawking 1993).

[2] Notably, Ayer later largely rejects logical positivism (Hanfling 2003). Indeed, much of logical positivism has been re-evaluated, reconsidered, misunderstood, and in some sense "softened" over the decades (Friedman 1999).

> that is tantamount to a reigning presumption of all science. And since science is always looking for simplifications, looking for the greatest generality it can muster, functionalism as a practice has a bias in favor of minimalism, of saying that less matters than one might have thought. (Dennett 2001, 39)

In Dennet's quote, we also see emphasis on how functionalism also relates to minimalism. The minimalism implied in Dennett's quote can be interpreted quite generally to suggest that a complete understanding of the fundamental constituents of reality does not really matter as much as our understanding of how reality acts. Deep conclusions about the fundamental essence of reality are not necessarily that important for a functionalist view of science and hence for scientific minimalism as characterized here. This relates to epistemological prudence noted previously in the sense that it is consistent with the idea that science should generally not make deep or broad ontological claims that cannot be adequately supported with empirical evidence. It may even be, as functionalism and epistemological prudence can be interpreted to imply, that the fundamental physical *essence* of the world is something that science cannot yet, or perhaps never, make confidence claims about.

4.4 Increasing Verisimilitude

Despite the general minimalism implied by epistemological prudence and functionalism, scientific minimalism still claims that physical reality is understandable or knowable to a large extent—while also acknowledging the possibility of fundamental limits to our knowledge about the physical world. The previous considerations regarding epistemological prudence and functionalism can still be consistent with the view that science is in some sense primarily about approximating the truth—or perhaps more strongly, and more importantly, that science is about increasingly approaching the truth over time—a goal for science that can be called *increasing verisimilitude*.

Scientific minimalism can include the idea that science generally approaches greater degrees of truth about the world without ever necessarily achieving complete truth. Some might argue that even this is a strong claim for a scientific *minimalist* perspective. Indeed, weaker versions of scientific minimalism do not necessarily have to adopt science as approximating truth, let alone doing so to an ever-increasing degree over time. However, this characterization is chosen here to help define scientific minimalism in a way that is more useful and revealing with respect to issues and worldviews that will be explored further in this book.

One end of a spectrum of thinking regarding the role of truth in science is that science pursues and achieves "complete truth." On the other end of the spectrum is essentially the position that is there is effectively no role for truth in scientific pursuits. Somewhere in between these extremes for the role of truth in science lie two models for thinking about the scientific enterprise: Larry Laudan's "problem-solving model" (Laudan 1977) and Philip Kitcher's "significant truth model" (Kitcher 1993).

Verisimilitude is often understood to connote the appearance of being true, often through what appear to be authentic details. The definition used here emphasizes that verisimilitude is considered to be an *actual* approximation of truth. Increasing verisimilitude says that the goal of science is to get closer and closer to the truth as theories and evidence improve over time. An obvious challenge is clarifying what is meant by "increasingly approximating the truth," and this will be addressed later. Assessments of Laudan's problem-solving model and Kitcher's significant truth model will help illuminate the potential role of increasing verisimilitude as a reasonable goal for science. While characterizations of truth can obviously be very complex and have a rich philosophical history, it is not our intent here to cover the important wide range of considerations regarding the fundamental nature of truth. Instead, the common notion of truth as "reality or actual existence" is sufficient and compatible with the use by Laudan and Kitcher.

The heart of Laudan's problem-solving model claims that the goal of science is to solve problems and that scientific progress is characterized by increasing problem-solving effectiveness. While Laudan does not deny the existence of truth, he claims that it is not logically connected to problem-solving effectiveness, and as a result, truth does not connect to judgments about progress in science.[3] Kitcher claims that the goal of science is to attain or discern significant truths about the world where significance is characterized as charting divisions and establishing explanatory dependencies. For Kitcher, truth plays a crucial role in that "significant truth" can be achieved.

Both Laudan and Kitcher recognize some relationship between science and truth. Laudan suggests that while truth is not necessarily connected to progress, his model does not preclude that possibility. Kitcher focuses on truth because his model makes significant truths accessible via the pursuits of scientific methods. Indeed, Kitcher writes: "if the uses I make of the concept of truth are incoherent, then my account of progress will be wrong" (Kitcher 1993, 90).

Laudan sees science as primarily a problem-solving activity, and he acknowledges progress in science as increasing problem-solving effectiveness. Laudan sees truth as disconnected from problem solving and hence from progress in science. However, what is important for our treatment here is that Laudan also acknowledges that his position does not necessarily preclude a role for truth. He writes:

> There is nothing in this model which rules out the possibility that, for all we know, scientific theories are true; equally, it does not preclude the possibility that scientific knowledge through time has moved closer and closer to the truth. Indeed, there is nothing I have said which would rule out a full-bodied, "realistic" interpretation of the scientific enterprise. But what I am suggesting is that we apparently do not have any way of knowing for sure (or even with some confidence) that science is true, or probable, or that it is getting closer to the truth. (Laudan 1977, 126)

---

[3]Laudan also articulates a reticulational goal for science in his (1984) book, *Science and Values: The Aims of Science and Their Role in Scientific Debate*, where there is arguably some conflict between his reticulational view and his problem-solving model discussed here. His problem-solving model is most relevant here because it is closest to increasing verisimilitude on the spectrum of ideas of the goals of science as they relate to the role of truth. For example, Doppelt (1986) provides an assessment of Laudan's reticulational goal, calling into question whether Laudan sufficiently defends forms of scientific relativism.

As Newton-Smith puts it, “Laudan is no atheist. He does not wish to deny the existence of truth” (Newton-Smith 1981, 185).

Kitcher holds the goal of science to be that of obtaining “significant truth.” He writes:

> Our primary tasks are to achieve a language that recognizes natural divisions and a set of explanatory schemata that pick out dependencies. To the extent that we can then advance true answers to significant questions, we aim to do so. But we know, on the basis of understanding our own cognitive limitations, that we will sometimes only be able to determine the approximate values of magnitudes, and that in dealing with complex systems we shall have to provide an idealized treatment if we are not to lose sight of the structure of the situation. (Kitcher 1993, 150)

With respect to achieving these truths, Kitcher further remarks:

> I have not simply contended that the sciences aim to arrive at concepts that single out natural kinds, schemata that correctly capture objective dependencies, significant statement that are true, but that various parts of the sciences achieve these aims. (Kitcher 1993, 150)

Kitcher puts himself close to increasing verisimilitude by noting how a completely true account of the world may not be possible: “The aim of arriving at a literally true story of the world might not only be something that is unattainable, but something we can recognize ourselves as being unable to attain.” (Kitcher 1993, 150). So, Kitcher claims the accessibility of significant truth while at the same time suggesting inaccessibility of literal truth.

A progression of unifying theories has allowed science to increasingly predict and explain more, and arguably with a smaller number of independent theories. But perhaps Kitcher goes too far in claiming they are “significant truths” as opposed to merely approximations of truth? Kitcher does refer to “approximate truth of universal generalizations” (Kitcher 1993, 122) which indicates a sensitivity to the notion of verisimilitude at least as it applies to universal generalizations. Further sensitivities of Kitcher to increasing verisimilitude can also be seen when he refers to verisimilitude as shorthand for a “complex of relations” (the use of “relations” is particularly relevant to the next chapter that explores a relationalist framework) that can get “closer to the truth” (Kitcher 1993, 123).

Increasing verisimilitude then seems to be a reasonable implication of the skeptical views of Laudan and more ambitious views of Kitcher in the sense that increasingly approximating the physical truth about the universe is plausibly within reach. Kitcher’s goal for science is arguably supported by what appears to be increasing explanatory and predictive power of science over time and does not necessarily require a commitment to “complete truth” as goal for science. Indeed, weaker forms of scientific realism lean heavily on the “no-miracles” argument which suggests it would be a miracle if our successful scientific theories were not at least approximately “true”. Related, structural realism suggests that while we may not know the true nature unobservable objects, we can have confidence in the mathematical and/or other kinds of structures (e.g. semantic) of successful scientific theories, in part because underlying structures can often be preserved in progressing from one theory to another (Worrall 1989, Suppe 1989).

4.5 Scientific Minimalism and Value Theory

While scientific minimalism can be interpreted to suggest that we can, and perhaps should, reserve judgments about ultimate philosophical questions of value (from which meaning, purpose, morality, ethics, etc., can follow), it is also reasonable to suggest that scientific minimalism can still support the idea that science can importantly inform explorations of value. This was touched on previously in the fact–value interplay discussion where biological evolution was used as an example to show how the facts of scientific knowledge can relate to, and arguably heavily influence or partially determine values. In general, however, what scientific minimalism would confidently say about human values is very limited. Scientific minimalism would suggest that many human values can, and probably should, be informed by scientific understanding (ranging from evolutionary biology to implications of cosmology), but that such interpretations should be considered to be highly provisional[4] and, in many cases, too speculative to have much confidence in.

But scientific minimalism would probably at least allow for the theoretical possibility that increasing empirical evidence, and hence greater epistemological confidence, could be obtained over time regarding human values (Harris 2010). But such empiricism as applied to human values would be very difficult, and perhaps impossible in some cases, to confidently obtain, and hence should be treated very carefully and viewed with a high level of healthy skepticism and concern for unintended consequences (Denning 2009). A broader form of scientific value theory could go further to expand the power of science to explain and influence values, but a scientific minimalistic position would suggest that applying scientific knowledge to deep questions about human values, meaning, purpose, ethics, etc., cannot be done with a high level of confidence with our present knowledge. Nevertheless, such pursuits may be helpful to pursue as we become more scientifically competent regarding human values—and perhaps value in a more general sense.

Scientific minimalism would then suggest that views such as those articulated by Teilhard de Chardin and Swimme and Berry that seem to imbue the universe with a deeper meaning and purpose, intrinsic or "objective" spirituality, or teleology of some kind, do not generally have scientific standing or sufficient epistemic credibility. Scientific minimalism would generally not be interested in such views as avenues of inquiry and would be very skeptical about the substance of such claims. Another

[4]Michael Shermer, in *The Science of Good and Evil* (2004), explores the idea of a "provisional morality" which is a straightforward common sense notion of moral sentiments, based largely on evolutionary predispositions that create moral norms which are accepted by most people most of the time under most conditions. Descartes' *Discourse on the Method of Rightly Conducting One's Reason and of Seeking Truth in the Sciences* also explored the idea of a "provisional morality," but without the knowledge of evolutionary forces that have shaped much of our moral sentiments.

way of thinking about some forms of scientific minimalism is Carl Sagan's caution that extraordinary claims require extraordinary evidence.

4.6 Quantum Theory: What to Make of It?

As explored previously, quantum theory is often used, perhaps overly used, to inform speculation about many questions ranging from cosmic origins and futures to consciousness and "spirit." Here, we will explore quantum theory in the context of scientific minimalism to help inform the extent to which quantum theory might be credibly extrapolated to broader speculative claims such as "retroactive observer participancy" noted in Chap. 2 as well as suggestions in subsequent chapters that leverage a relational framework for thinking about cosmological worldviews.

Of particular interest for interpretations of quantum mechanics that will be explored here are the suggestions touched on earlier by James Gardner and Paul Davies that observers, or consciousness more generally, may play a role in bringing the universe into existence via a limited form of "retrocausation." These speculations are based on original conjectures of retroactive observer-participancy from John Wheeler (as previously touched on in Chap. 2). Perhaps most provocatively, these speculations include the idea that observers may exert influence on the emergence of the universe's physical constants and the laws of physics themselves—something that Davies particularly focuses on and admits is a highly speculative, but perhaps at least plausible theoretical possibility.

Wheeler's idea of retroactive observer-participancy is largely derived from original conceptions of quantum mechanics that were articulated by the Copenhagen interpretation—a major part of which emphasized the importance of observation, or measurement, to create a discernible state of reality. So, this section will focus heavily on examining the Copenhagen interpretation using assumptions and features of scientific minimalism touched on previously. This can help assess the plausibility of retroactive observer-participancy and hence the plausibility that life and intelligence might ultimately play a role in bringing our specific universe into existence. We will also briefly explore other relevant interpretations of quantum mechanics that relate to scientific minimalism as well as to subsequent topics that will be explored in following chapters.

4.6.1 Basic Quantum Mechanics

When Max Planck proposed the idea of the quanta in 1900 to explain black-body radiation, he initiated one of the most successful scientific adventures in history. That endeavor has included much discussion and debate about what quantum mechanics implies about our world and our ability to know it. While most scientists agree on the mathematical formalism of quantum mechanics and acknowledge its tremendous predictive and explanatory power, there remains a wide range of interpretations regarding its broader implications. Rather than completely defining what is meant by an interpretation of quantum mechanics, this treatment will primarily examine one of the most influential interpretations, namely the Copenhagen interpretation, along with a few other relevant interpretations such as Bohmian mechanics and relational quantum mechanics (RQM), from which an understanding of what an interpretation is, or should be, will hopefully become more apparent.

In *Quantum Mechanics and Experience*, David Albert spells out quantum mechanics in terms of five basic principles: (1) physical states, (2) measurable properties, (3) dynamics, (4) probability, and (5) collapse (Albert 1992). With the exception of the notion of collapse, this outline of quantum mechanics features will serve us well for establishing the essential workings of quantum mechanics. The notion of collapse belongs more to the realm of interpretation than that of quantum mechanical formalism and will be discussed later. There will be a mix of references to wave mechanics and matrix mechanics because each lends itself better depending on the context. Later, we will touch on the relevance of quantum field theory—a more modern incarnation of quantum mechanics.

Physical states, in Albert's five principles noted above, are understood to represent the physical situation or physical configuration of a system. Physical states are represented by vectors or state vectors. The total of all possible states of a system, or of all the state vectors for that system, represents the system's associated space or vector space. The state vectors pick out some particular state of the system.

Measurable properties are understood as observables of a physical system and are represented by linear operators on the vector spaces associated with the system. If the new vector generated by the operator is in the same direction of the original vector, the vector is said to be an eigenvector of the operator and the value of proportionality is said to be an eigenvalue of the operator. Collectively, the system is said to be in an eigenstate of the property in question and the corresponding eigenvalue is the particular value of the measurable property.

The dynamics of a physical system is concerned with how the system state vector (or wave function) evolves over time given the forces and constraints to which the system is subjected. The dynamics of a state vector is usually cast in the form of a differential equation of motion called the Schrödinger equation whose solution is a wave function, which is the time evolution (or dynamics) of the system. In terms of vectors, this equation of motion represents changes in vector direction, not length.

Probability is concerned with the connection to experiment—namely the results of measurements. The issue here is to understand what happens if we measure a system when it is not in an eigenstate. The value of the measurable property will not be known to us. Quantum mechanics tells us that the outcome of such a measurement is a matter of probability. Specifically, the probability is given by the square of the wave function.

The above approach can be used in the following way to predict the behavior of a particular physical system: (1) Identify the vector space of the system, (2) specify the correspondences of measurable properties to operators, (3) specify the correspondences of physical states to vectors, (4) determine the state vector via measurement, (5) calculate the time evolution of the state vector via the Schrödinger equation, and (6) calculate the probabilities of particular outcomes of future measurements. Steps 5 and 6 can be repeated as necessary.

4.6.2 Conceptual Problems: Probability and Classical Physics

One essential challenge of quantum mechanics is its apparent probabilistic or indeterminate features. The probabilistic aspect of quantum mechanics can be associated with Max Born's initial interpretation of the wave function, and "indeterminacy" is associated with Heisenberg's uncertainty principles. However, each term captures the essential difficulty with quantum mechanics, namely that it makes statistical/probabilistic predictions. To understand the difficulty this has presented, it is important to briefly cover the historical context by which the probabilistic nature of quantum mechanics was born. That historical context will help us understand what is meant by the probabilistic nature of quantum mechanics. A second related problem of quantum mechanics is its relation to classical physics and whether the two can be reconciled, including the challenges of the statistical and probabilistic nature of quantum mechanics just noted (Bub 1974).

By some accounts, the wave function in the Schrödinger formulation was considered by Schrödinger to be an element of physical reality (Jammer 1974). This view implied a kind of undulatory conception of reality based on classical physics. The wave packet was essentially a particle in this view. But this perspective had problems, two of which were: (1) the dispersion of the wave packet and hence the impermanence of the "particle" and (2) the discontinuous change (or "collapse") that the wave function undergoes during a measurement. Both of those concepts have implications for probability and indeterminism as well as for a reconciliation with classical physics.

The scepter of probability was raised by Max Born in his struggle to reconcile the wave mechanical formulation of the Schrödinger equation with the corpuscular nature of electromagnetic radiation that was being supported by experiments. For a corpuscular entity such as an electron, the wave function could only represent a probability for the location of the electron. Born said, "The motion of particles conforms to the laws of probability, but the probability itself is propagated in accordance with the law of causality" (Jammer 1974, 40). Born's probabilistic interpretation went a long way to address the concerns over the dispersion effect because the wave function was not thought of as a physical entity, but only a mathematical tool yielding probabilities. The issue of the collapse of the wave function however was not adequately addressed by Born's interpretation. Nor did his interpretation adequately address the apparent wave phenomena exhibited by particles.

The problem of probability and indeterminism was made larger by Werner Heisenberg in 1929 when he formulated his indeterminacy relations. Heisenberg attempted to solve the following problem: Matrix mechanics did not define the concept of a path or orbit of an electron, and wave mechanics implied dispersion of the path of the wave packet so that the deterministic path observed in cloud chambers could not be reconciled with either formalism. Recalling a conversation with Einstein, Heisenberg remembered that Einstein proffered the view that "it is the theory which decides what we can observe" (Jammer 1974, 57). If Heisenberg could show that the theory denies the strict observability of the trajectory of a particle, a connection could be made between the mathematical formalism and the experimental evidence. What resulted were the Heisenberg indeterminacy relations where variables like position and momentum could not simultaneously be known to complete accuracy. Heisenberg attributed the probabilistic nature of quantum mechanics to this indeterminacy. Heisenberg, however, did not see his indeterminacy relations as sufficing for an interpretation of the wave mechanical formalism of quantum mechanics. On this matter, he was persuaded by Bohr who said that "recourse to both, the particle and wave pictures are needed to determine in all instances the limits to which classical concepts are applicable" (Jammer 1974, 59).

Another significant concern that plagued Bohr and others that relates to the problems of probability and measurement was the issue of how to reconcile quantum mechanics with classical physics. For Bohr, quantum theory was more about mathematics than it was about physics, and this concerned Bohr—a bridge between the two was necessary for a full understanding of reality. Bohr saw classical concepts as indispensable to quantum mechanics. He wrote:

> …however far the phenomena transcend the scope of classical explanation, the account of all evidence must be expressed in classical terms. The argument is simply that by the word 'experiment' we refer to a situation where we can tell others what we have done and what we have learned and that, therefore, the account of the experimental arrangement and of the results of observations must be expressed in unambiguous language with suitable application of the terminology of classical physics. (Gibbins 1987, 54)

4.6.3 The Copenhagen Interpretation

The Copenhagen interpretation of quantum theory is arguably best understood more as a philosophy of science as opposed to a strict interpretation of quantum mechanics. It was put forth by Bohr largely to interpret quantum mechanics within the broader context of science, metaphysics, and epistemology. Max Jammer's historical treatment of the Copenhagen interpretation sets the stage of the 1927 International Congress of Physics and cites a direct quote from Bohr on his intention for his talk:

> On the one hand, the definition of the state of a physical system, as ordinarily understood, claims the elimination of all external disturbances. But in that case, according to the quantum postulate, any observation will be impossible, and, above all, the concepts of space

> and time lose their immediate sense. On the other hand, if in order to make observation possible we permit certain interactions with suitable agencies of measurement, not belonging to the system, an unambiguous definition of the state of the system is naturally no longer possible, and there can be no question of causality in the ordinary sense of the word. (Jammer 1974, 87)

Bohr further stated: "An independent reality in the ordinary physical sense can neither be ascribed to the phenomena nor to the agencies of observations" (Jammer 1974, 91). However, as indicated in the section outlining challenges in quantum mechanics, Bohr was concerned with interpreting quantum mechanics in terms of classical concepts. For Bohr, this was essential for understanding quantum mechanics. As a result, a further element of Bohr's Copenhagen interpretation lies in his use of the superposition principle to identify wave packets with particles.

With respect to the indeterminacy relations, we see Bohr differing from Heisenberg because while Heisenberg was satisfied using both particle and wave descriptions, Bohr saw a need for revising the classical conception of explanation. That is, Bohr's view was really more of an epistemological concept than an ontological one. Roland Omnes tellingly refers to it as the "Copenhagen Epistemology" (Omnes 1994, 85)—it was a philosophical notion about modes of description. Modes of description could complement each other. For Bohr, the collapse of the wave function was what caused a switch from one mode of description to its alternative complementary mode.

The Copenhagen interpretation cannot be adequately understood without briefly contrasting it with Einstein's view since this was a central debate in understanding quantum mechanics. Einstein articulated his view at the Fifth Solvay Congress in October of 1927 when he outlined two viewpoints with respect to the probabilistic nature of quantum mechanics. Viewpoint 1, or the ensemble interpretation, holds that the Schrödinger waves do not represent one individual particle. Instead, they represent an ensemble of particles distributed in space. In this view, the square of the wave function can then be seen to be representing the probability density that there exists some particle of the ensemble. Viewpoint 2 treats quantum mechanics as a complete theory of individual processes where a wave packet represents a particle.

The "EPR" argument, or paradox, based on thought experiments from Einstein, Podolsky, and Rosen captures Einstein's essential discomfort with quantum mechanics. To Einstein, quantum mechanics was incomplete (Einstein et al. 1935). If quantum mechanics was about ensembles, then it provided only statistical knowledge. If quantum mechanics was about individual objects, then a "spooky" non-local effect was needed. Either way, quantum mechanics was incomplete. However, it should be noted that Peter Gibbins further explains Einstein's concerns by noting that for Einstein, the collapse of the wave function would require some non-local force which would have to act faster than the speed of light. So not only does Einstein perceive quantum mechanics as incomplete, but he also sees it as a contradiction of his own theory of relativity (Gibbins 1987). For Einstein, probability then ultimately implies non-locality, and unless non-locality can be explained, non-locality implies incompleteness.

There were later versions associated with the Copenhagen interpretation, largely in response to the EPR argument. The EPR argument outlined an epistemological criterion for physical reality, which called for an intrinsic set of properties independent of the observer (Mermin 1991). Bohr defended his position by refuting the EPR criterion for physical reality. He claimed that the "object under observation and observing apparatus from a single indivisible system is not susceptible to any further analysis, at the quantum level, into separate distinct parts" (Jammer 1974, 197). This is an element of holism which can also be considered an important element of the Copenhagen interpretation and will be touched on further in subsequent chapters.

Richard Healey (1989) explores weak and strong versions of the Copenhagen interpretation. The weak version incorporates the "collapse" of the wave function or the "projection postulate" as the source of values for dynamical variables. In this view, quantum mechanics is about individual systems. The strong view says that quantum states should be assigned to "ensembles" or a set of similar individual systems and that the behavior of such collections is predicted in terms of statistics of experimental results.

In Healey's weak interpretation, the Heisenberg indeterminacy relations imply that there is a limitation to the precision with which position and momentum are simultaneously defined on any individual system. Healey points out that the "measurement problem" is a significant problem for proponents of this view because they must show how measurement affects the system so as to yield the result determined by the experiment. The strong view implies that quantum mechanics has nothing to say about the dynamical properties of the quantum system at a time it is not being observed. Healey sees the strong view as being closer to Bohr's views and points out that the strong view requires describing and understanding the measuring apparatus classically.

One way to assess the Copenhagen interpretation is that it adequately deals with indeterminacy by simply stipulating that indeterminacy is a feature of the world in some fundamental sense. Indeterminacy is simply part of the quantum mechanical nature of the world. In that sense, it does not need interpreting—however unsettling it may be. Where Heisenberg saw indeterminacy as a physical aspect of the world, Bohr saw it as an epistemological issue—a reason to revisit the classical conception of explanation. Regardless, we can say the Copenhagen interpretation acknowledges an unavoidable role for probabilities. The probabilistic feature of quantum mechanics is not really in question, but the exact role and broader implications for probability is. Without stronger claims, the Copenhagen interpretation is diluted to the point of arguably being unhelpful with respect to understanding the probabilistic feature of quantum mechanics. Here, we see a dizzying blurring of many considerations to the point where it is very difficult to assess whether or not the Copenhagen interpretation adequately addresses the probabilistic features of quantum mechanics.

And while holism can be considered an important, and perhaps ultimately useful, aspect of the Copenhagen interpretation (which will be discussed more later), the validity and usefulness of the Copenhagen interpretation seem to depend on how it is defined and on how the problems are defined. On the whole, it appears to be very difficult, if not impossible, to make a convincing argument for the validity or usefulness of the Copenhagen interpretation as an interpretive framework. It is somewhat curious then that the Copenhagen interpretation has endured to this day with apparent influence (Schlosshauer et al. 2013). David Albert suggests that the "standard" way of thinking about quantum mechanics can be traced back to Bohr and the Copenhagen interpretation (Albert 1992). Healey refers to the Copenhagen view as orthodox (Healey 1989). Some may suggest that the Copenhagen interpretation is the best we can do for now, but there are alternatives.

4.6.4 *Ensemble/Statistical and Instrumentalist Interpretations*

Perhaps the most conservative interpretation of quantum mechanics is the "ensemble" or "statistical" interpretation. Its distinction from Copenhagen interpretations is subtle, but important. The interpretation can be traced back to the beginning of the development of quantum mechanics when Max Born emphasized the statistical nature of the emerging quantum formalism. As noted previously, this interpretation suggests the formalism of the wave function applies to a collection of similar particles and systems, not to an individual particle which otherwise creates the difficulties that arise in the Copenhagen interpretation. Einstein wrote:

> The attempt to conceive the quantum-theoretical description as the complete description of the individual systems leads to unnatural theoretical interpretations, which become immediately unnecessary if one accepts the interpretation that the description refers to ensembles of systems and not to individual systems. (Einstein 1949, 671)

Importantly, as opposed to the Copenhagen interpretation, the ensemble interpretation is essentially agnostic about determinism, the realism of quantum phenomena, and whether quantum mechanics is a complete theory or not. The ensemble interpretation merely provides a statistically robust explanation for what the wave function is telling us, and this seems to avoid the interpretative difficulties of quantum mechanics seen previously with the Copenhagen interpretation (Ballentine 1970).

The ensemble interpretation can be said to be consistent with what might be considered an even more minimalist interpretation, namely an instrumentalist perspective. An instrumentalist perspective is a general scientific methodological orientation (not really an interpretation per se) that simply acknowledges repeatable relationships between formalism and experimental practices, i.e., measurements. On the ensemble interpretation, quantum mechanics is merely a tool for determining the statistical properties of a probability distribution.

4.6.5 Hidden Variables and Bohmian Mechanics

If the Copenhagen interpretation of quantum mechanics is at one end of the spectrum, where determinism is fundamentally in question, then at the other end of the spectrum is something like the de Broglie–Bohm theory (hereafter referred to as Bohmian mechanics) which is a deterministic quantum mechanical theory (Durr et al. 1996). Bohmian mechanics can be thought of as part of the "hidden-variable" program that emerged early on in the development of quantum theory. The hidden-variable interpretation is based on the view that quantum theory is incomplete and therefore needed additional variables to be deterministic and complete.

Bohmian mechanics does not require that an observer "collapses" the wave function in order for a reality to materialize and be measured (Maudlin 1995). Instead, in Bohmian mechanics, there are both waves and particles, where the wave, often called a "pilot wave," guides the position and trajectory of particles. Bohmian mechanics predicts definite positions of particles and uses the Schrödinger wave function as the pilot wave. It is notable that while an observer can objectively determine a particle's position at any given time, the complete trajectory cannot be objectively determined in the sense that measuring the particle at any given time subsequently modifies the particle's trajectory, consistent with Heisenberg's uncertainty principle.

4.6.6 Objective Collapse

Objective collapse theories suggest that a real wave with an associated wave function does indeed collapse, but does so randomly and spontaneously or when a particular physical threshold is reached, independent of measurements by observers. The Ghirardi–Rimini–Weber (GRW) theory modifies the Schrödinger equation and posits a low-rate spontaneous probabilistic collapse within the time evolution of the wave function (Ghirardi et al. 1985, 1986).

Interestingly, it is suggested that the rate of these spontaneous collapses for an individual particle is something like once every hundred million years. However, because macroscopic objects like measuring devices consist of so many more particles than the individual particle collapse rate, there is always an extremely high probability that the entangled particle–device system will be seen in a collapsed state. That is, one of the particles within the experimental setup is very likely to have been collapsed, hence showing the entire system as collapsed and creating the appearance of collapse during the measurement. The particle is collapsed because of its *relationship or entanglement* with the much larger physical experimental system that is already collapsed, not necessarily because the observer creates the collapse through the act of observation.

Roger Penrose has suggested a different kind of objective collapse that is separate from the wave function which remains unchanged. This additional collapse is triggered by a small gravitational effect linked to Einstein's general relativity space-time (Penrose 2014).

4.6.7 Many-Worlds, Parallel Worlds

The often called "many-worlds" interpretation of quantum mechanics, perhaps better described as "parallel worlds", was first proposed by Hugh Everett in 1957 (Everett 1957) and, notable for the purposes of the relational framework to be developed in the next chapter, was more specifically referred to as a "relative state formulation." Everett referred to it as a metatheory with respect to numerous quantum theories and interpretations (such as the Copenhagen interpretation). The relative state formulation essentially says that completely independent states of observer (including physical instrumentation) and observed are not meaningful since both are correlated or "entangled" by observation/measurement, so the state of each must be specified and understood relative to each other—and the correlation results from observations/measurements between subject and object.

If the object is in a superposition (a combination of states—which is an assumed principle in most quantum theory formulations), then the entanglement of the object with the subject (observer or apparatus) during observation/measurement means that the subject then also goes into a superposition of states that correlate with the superposition states of the object. That is, the subject sees the object in both of its states (e.g., spin-up and spin-down). This has implied to Everett and others that there is a world in which the subject sees the spin as up and a world in which the subject sees the spin down—there are two worlds that result from the observation/measurement. The universe has essentially "split" into two different realities. In both branches, both worlds are equally real if we are to believe the deterministic time evolution formalism of the Schrödinger equation.

This relative state linkage between subject and object is important because when the deterministic Schrödinger equation is applied to the superposition of the combined subject–object system, Everett says that because the system is interlinked, there are then subsequent unitary evolutions of the linked, or entangled, systems. This arguably then removes the need for an actual wave function "collapse" since no collapse is required. It may appear to any observer in one branch that there was something like a "collapse" to create an actual determinate state, but in reality it is simply one part of the ever-evolving complex superposition of the overall universal wave function that is being observed. This can be interpreted as a picture of a universe that perpetually exists in a combination of many states—implying a kind of universal wave function of immense complexity and perhaps infinite variety—an implication that can be tied to our exploration of worldviews to come.

4.6.8 Relational Quantum Mechanics

A more recent view of quantum mechanics that was developed in the mid-1990s (e.g., Rovelli 1996) that has much consistency and overlap with most of the views already touched on is an explicitly "relational" interpretation, or what has been called "relational quantum mechanics" (RQM). Similar to the many-worlds formulation just covered, RQM emphasizes that the state of a system describes how the observed system (object) and the observer (subject) are correlated, i.e., related.

But critically, RQM is more general in the sense that all physical interactions are treated with a relative specification irrespective of the nature of the objects involved. Any normal physical interaction is in some sense the functional equivalent of a "measurement" in the sense that *any interaction* determines the reality that is thought to emerge from subject–object measurements of other interpretations. So, the theory is not about objects and subjects per se, but is rather about *how all physical objects interact with and relate to each other*. The theory is about the relationships between all objects, not the objects in and of themselves. In this sense, RQM can be seen as a more general form of the relative state formulation of many-worlds which appears to be primarily constrained to subject–object relationships where the subject is essentially restricted to an act of observation.

4.7 Scientific Minimalism and Quantum Theory

Quantum mechanics is one of the most important scientific developments in human history, and it will critically inform our exploration of the relationist framework in Chap. 5 and the cosmological worldviews that follow. But as explored previously, it is unclear how much explanatory power quantum mechanics has to inform foundational questions about the objective nature of reality, let alone philsophical questions of value and meaning. An interesting question is how important it is to fully explain the fundamental nature of the world anyway—a question that will be explored briefly here. We will apply scientific minimalism to various views of quantum theory to see what it might say about one of history's most challenging theories and the broader questions that quantum theory may relate to.

The various views of quantum mechanics previously explored all have features that are consistent with scientific minimalism. Nevertheless, there are important differences and varying degrees of consistency worth exploring. We can apply the three main characteristics of scientific minimalism noted earlier to the various views of quantum mechanics to see how consistent they are with a scientific minimalistic perspective. The three characteristics we will apply that were explained prior are epistemological prudence, functionalism, and increasing verisimilitude.

After surveying the implications of scientific minimalism for various perspectives on quantum theory, we will be in a better position to make an informed assessment regarding any role observers play in quantum theory. We should also be in a better position to develop and assess the broader cosmological worldviews in subsequent chapters, for which quantum theory is an important factor.

4.7.1 *Copenhagen Interpretations*

In the Copenhagen interpretation (CI) assessment we made prior, it was concluded that while much of the validity of the interpretation depends on how it is defined and how the problems are defined, it is nevertheless difficult to see how most, if not all, of the various forms of the arguments can sufficiently support the validity of just about any version of the Copenhagen interpretation. If we apply epistemological prudence in the form of Occam's razor, it is reasonable to think that wave function collapse caused by an observer from a previous state of superposition (an indeterminate combined state) is a significant assumption. It is significant not only because it lacks clear details about timing and mechanism, but also because it can be interpreted to assume the world does not exist in a definite state until it is observed. The Copenhagen interpretation appears to reject determinism on insufficient grounds.

However, CI is arguably moderately functional in the sense that it provides some form of a broader interpretative context for understanding potentially important implications of quantum mechanics. CI provides a helpful conceptual framework that also tends to focus more on system behavior and not what the system is fundamentally made of per se—a second element of functionalism noted earlier.

Regarding increasing verisimilitude, this is a little more difficult to assess for CI. If we take truth to mean knowing the actual physical state of the world along with fundamental details about how the world is made, it would appear that CI is far from that standard. Indeed, Bohr is reputed to have often noted, in opposition to Einstein, that the quantum revolution teaches us that the scientific endeavor is about what can be *said* about the world, not what *is*. Given this kind of epistemological orientation, it appears that CI only very weakly contributes to satisfying increasing verisimilitude as a goal of science. In fact, CI could be interpreted to be essentially agnostic about increasing verisimilitude as a goal for science.

Hence, as scientific minimalism was defined earlier, it appears CI meets only one of the three standards at a moderate level, i.e., the functionalism standard. It should also be noted that while many in the scientific community might find CI to be functional enough for various purposes, others do not. It depends in part on what functions a particular scientist is interested in, and this will hopefully become clearer as this analysis unfolds.

4.7.2 *Ensemble Interpretations*

It was noted previously that ensemble interpretations of quantum mechanics are essentially agnostic about determinism, the realism of specific quantum phenomena, and the completeness of quantum mechanics as a theory. This level of agnosticism makes ensemble interpretations very epistemologically prudent. The statistical nature of ensemble interpretations also makes it epistemologically prudent via Occam's razor because claims about specific conceptual mechanisms for dealing with individual particles are not needed, hence avoiding constructs like wave function collapse and a critical role for an observer. In ensemble interpretations, quantum mechanics is seen primarily as a tool for understanding probability distributions, nothing more.

Ensemble interpretations are also highly functional in the sense that such interpretations are consistent with most, if not all, empirical scientific practice. It was noted earlier that ensemble interpretations are consistent with instrumentalist perspectives in the sense that quantum mechanics provides formalism that allows for repeatable relationships between experiment and formalism. This is both functional and epistemologically prudent.

Regarding increasing verisimilitude, it can be argued that ensemble interpretations are more aligned with increasing approximations of truth than are Copenhagen interpretations, in part because there is a level of openness for further deeper scientific pursuits of microscopic phenomena. Also, because ensemble interpretations do not rely on an observer to create a physical reality, this suggests a level of commitment to an objective reality that can increasingly be better understood as our knowledge increases. Statistical/ensemble interpretations, then, appear to fare well in a scientific minimalist context.

4.7.3 Bohmian Mechanics

Bohmian mechanics may be seen as epistemologically prudent in the sense that it does not require a fundamentally mysterious role or mechanism by which an observer creates a real physical measurable state. However, Bohmian mechanics does introduce a new physical entity to quantum theory—i.e., a "pilot wave" or "guiding wave" which it assumes to be physically real. In this sense, Bohmian mechanics can be said to be less epistemologically prudent than say, ensemble interpretations, because a fairly significant physical entity is introduced and assumed to be physically real. Indeed, claiming the Schrödinger equation is a real physical entity is a fairly bold assertion for which scientific minimalism would question whether the physical evidence adequately supports that claim—although evidence such as wave interference patterns in double-slit experiments can be interpreted as physical evidence for such a construct.

Bohmian mechanics does appear to be highly functional and fairly consistent with experimental results of standard quantum mechanics. It offers an additional functional construct of an ontological wave that is a fundamental driver of particle dynamics. This additional element appears to provide at least theoretical functionality for many physicists and offers the possibility that an underlying reality can be further explored and understood, pointing to the possibility for increasing verisimilitude.

Expanding on the previous point about an additional functional construct of a guiding wave, Bohmian mechanics certainly seems to be committed to the idea of increasing verisimilitude as a goal of fundamental physical theories. Indeed, it was developed largely based on concerns of the apparent incompleteness of quantum mechanics, suggesting openness to the evolution of scientific theories to better explain the reality of what is essentially a deterministic world. Bohmian mechanics may in fact ultimately be contributing to that kind of increasing approximation of the truth by positing a real wave phenomenon that may in fact exist independent of observers. However, scientific minimalism as defined does not fully commit to (or is at least agnostic about) determinism and whether the world can be completely known, so in this sense Bohmian mechanics may overshoot expectations of scientific minimalism and hence raise questions about whether such a motivation leads to creating unnecessary theoretical constructs such as a pilot wave.

Related to the last point noted above, another implication of Bohmian mechanics that one might hope a realist intuitive deterministic theory could counter is that it is explicitly non-local, as are both Copenhagen and ensemble interpretations. This non-locality seems to imply some kind of "spooky action at a distance" or, perhaps more generically and more accurately, a "correlation" at a distance. Non-locality is a significant challenge for many physicists given a lack of mechanism for understanding how such an unusual and counterintuitive phenomenon could work. A commitment to some form of non-locality would also seem to reduce epistemological prudence since it assumes an apparently dogmatic commitment to an unusual physical phenomenon without sufficient evidence.

Bohmian mechanics might be said to be somewhere between moderately and highly minimalist. It can be considered highly functional and consistent with increasing verisimilitude, but it is not highly epistemologically prudent since it posits non-locality and a significant physical entity in the form of a something like a "pilot wave."

4.7.4 Objective Collapse

By suggesting that there is indeed a wave function collapse that occurs in nature independently of observers and measurements, objective collapse theories do exercise a kind of epistemological prudence, because, as in the case with Bohmian mechanics, the theory does not require an observer to create a real measurable state via wave function collapse triggered by observation. However, the two objective collapse theories explored previously do nevertheless introduce complex and essentially random physical mechanisms that trigger collapse. These mechanisms appear to be detectable in principle, but have not yet had sufficient experimental verification.

The lack of commitment to determinism makes objective collapse theories epistemologically prudent in the sense that they do not rely on determinism. However, these theories do seem to say that nature is both random and determinate. The apparent ontological claim regarding random states of nature suggests that clear deterministic physical laws are not at work at all times, and this arguably raises questions for epistemological prudence since it seems to violate expectations of consistency. However, assuming ontological randomness in nature also has a certain parsimonious appeal—certainly an epistemological parsimony if not an ontological parsimony. On balance, given that these theories provide predictable theoretical, but as of yet unverified physical mechanisms for an objective collapse of a real wave function, we can think of objective collapse theories as moderately epistemologically prudent.

Objective collapse theories are functional in the sense that they apparently make most of the same predictions of standard quantum mechanics. But importantly, it is claimed that these theories make different predictions under certain circumstances that seem to be experimentally within reach. This has promise for increased functionalism because this can allow for some direct comparison with standard views of quantum mechanics if and when such empirical evidence is obtained. However, as emphasized previously, as of right now this is a theoretical prospect—one that may or may not ultimately make an important difference. On balance then, it might be said that objective collapse theories have a moderate level of functionalism.

As with Bohmian mechanics, objective collapse theories are committed to the objective truth-seeking endeavor of science in the sense that physical mechanisms are proposed and are, in principle, testable. Objective collapse theories make an attempt to build on and modify standard quantum mechanics to better explain an independent physical world. Objective collapse theories can be interpreted as tacitly acknowledging that there may be some limitations to what science can ultimately know and predict, while at the same time increasingly approximating truth about the physical world by considering new ideas and testing them, however subtle and complex they may be. On balance then, objective collapse theories might be said to be moderately minimalist.

4.7.5 Many-Worlds

The many-worlds, or "parallel worlds," interpretation, based on the original relative state formulation of Everett, is certainly an intuitively challenging concept—at least as far as the apparent ontological implications are concerned. Positing the existence of actual parallel universes (potentially an infinite number of them) resulting from interactions with measuring devices and/or observers is a significant implication to say the least—perhaps the most significant scientific claim ever made. Indeed, the unusual nature of these claims makes a scientifically minimalist assessment difficult.

The many-worlds interpretation may be thought of as epistemologically prudent in the sense that quantum mechanics is assumed to be a complete theory and so does not require any additions or modifications. The interpretation is realist, deterministic, and claims to be observer-independent, therefore not requiring an observer-driven wave function collapse for there to be a reality. Instead, the particular state picked out by the relative state of the subject–object entanglement (which happens during a subject–object interaction) simply continues on. But notice that these realities subsequently created by a subject–object entanglement/interaction are not measurable, apparently not even in principle. So essentially, the claim is that an unmeasurable reality is created. This calls into question whether many-worlds is a scientifically epistemologically prudent theory. It appears it is not, because not only do the implications seem unverifiable, but they appear to be unverifiable in principle.

Also, for the many-worlds interpretation to work, for there to be a lack of wave function collapse, it must invoke a universal wave function that never collapses, but simply constantly branches instead. Positing a universal wave function might be considered epistemologically prudent in the sense that it simply results from applying what is assumed to be the complete theory of quantum mechanics to the universe as a whole. However, there are two important tightly linked assumptions here. First, quantum mechanics must be assumed to be a complete theory, and second, that it can indeed be applied to the universe as a whole. It is questionable whether being a "complete theory" necessarily implies that the theory can indeed be applied to the universe as a whole single entity, as opposed to being applied to everything within the universe.

It appears that while many-worlds does not require any modifications to quantum theory and can eliminate the need for wave function collapse, it nevertheless makes a fundamental and highly questionable assumption about the completeness of quantum mechanics and its application to the universe as a whole—which itself invokes another theoretical construct, i.e., a universal wave function. Equally important, the implications of the many-worlds view appears to be fundamentally untestable. If claims about logically necessary conclusions of the existence of many parallel universes are based on questionable premises to begin with and have significant untestable implications, this can be interpreted to suggest that the many-worlds interpretation is weak on epistemological prudence.

Because the many-worlds theory appears to be untestable, even in principle, its functional value is also arguably low. Also, it is not clear what additional significant functional value the interpretation provides in its full-blown form. The narrower "relative state formulation" originally invoked by Everett is a part of the overall view that does appear to be useful, as will be explored in the next assessment of relational quantum mechanics. However, taken as whole, including implications of infinite untestable parallel universes, it appears that the many-worlds interpretation is not particularly functional for scientists on a day-to-day operational working level. Nevertheless, theoretical functional benefits of many-worlds are something many continue to explore (Hartle 1968; Deutsch 1997; Saunders et al. 2010; Wallace 2012), although as noted previously, relative state constructs do seem to have functional value for other related considerations noted prior.

While there is clearly an assumption of an objective truth to be known, because many-worlds is thought to be essentially a complete and determinate theory of quantum mechanics, it arguably overshoots the increasing verisimilitude suggested by the kind of scientific minimalism explored here. Also, every parallel universe is apparently determined and fully knowable only within that particular universe. This is a strong claim that scientific minimalism could be open to in principle if the evidence were compelling enough to demonstrate it. However, the bar for such evidence is high (e.g., completeness of quantum mechanics, actual measurable physical existence of parallel universes, etc.), and the many-worlds interpretation seems to fall well short of that. Based on the above reasoning, the many-worlds interpretation appears not to be scientifically minimalist.

4.7.6 Relational Quantum Mechanics

Relational quantum mechanics (RQM) is highly epistemologically prudent in the sense that it makes very few assumptions and arguably makes few bold claims. RQM assumes that quantum mechanics is complete and does not require additions to the theory, nor does it have any unusual theoretical constructs or ontological implications.

The knowledge claims of RQM are moderate in the sense that it articulates specific knowledge about the physical world with respect to frames of references, just as in special relativity. Those frames of references do not belong to any special set of circumstances or objects. That is, there is no distinction between microscopic and macroscopic objects or phenomena—making it importantly different from Copenhagen interpretations. Indeed, there is no universal wave function as in Bohmian mechanics or many-worlds. Locality versus non-locality is essentially irrelevant in RQM because everything is relevant to local frames of references, so effectively, locality is conserved—at least in practical terms from any given observer's point of view.

Similar to the many-worlds interpretation, the relational interpretation is perhaps only moderately functional in the sense of offering specific new working prescriptions or implications for practicing quantum physicists. It does however offer an important theoretical context within which to address a number of challenges regarding the interpretation of quantum mechanics. In this sense, RQM is theoretically useful while not needing to add any new theoretical constructs or having unusual ontological implications. It could be argued that its functionality is decreased by the limitation that it cannot be applied to the universe as a whole.

RQM is consistent with increasing verisimilitude in the sense that it acknowledges fundamental limits to what can be objectively known by many observers simultaneously since it claims all knowledge is relative to local frames of reference and so can be different for different observers. This may or may not be an epistemological limitation, but it appears that RQM would claim it is a fundamental limitation.

The increasing verisimilitude of scientific minimalism is essentially agnostic about whether there are truly fundamental limits to what we can know about the world, but RQM can be interpreted to suggest there are likely to be fundamental limits to our knowledge. However, a more practical version of increasing verisimilitude for scientific minimalism can be interpreted to suggest that RQM prematurely implies that limitation. That premature articulation may arise from a starting assumption that quantum mechanics is a complete theory—a concern that has been raised previously a number of times in assessing other interpretations of quantum mechanics.

It appears that while RQM is highly epistemologically prudent, it is only moderately functional while being sympathetic to increasing verisimilitude as a plausible goal for science. The high epistemic prudence, along with some moderate functionality and consistency with increasing verisimilitude, makes RQM reasonably consistent with scientific minimalism.

4.7.7 Thematic Summary of Scientific Minimalism for Quantum Theory

All versions and interpretations of quantum mechanics previously explored are basically scientifically and operationally functional to some extent (with the possible exception of many-worlds which may have high theoretical functionality), perhaps hinting that functionalism is what matters most for a minimalist scientific worldview. Indeed, many physicists are not particularly concerned about broader interpretations of quantum mechanics which is sometimes summed up as "Shut up and calculate!"[5] Many physicists see quantum mechanics primarily as a tool for operational scientific practices and technological advances. Some see it as a tacit acknowledgment, or at least as gesturing at the possibility, that fundamental reality might be out of reach—that quantum theory tells us something about how we are epistemologically limited and related to the world. In general, an acknowledgment of fundamental limitations to our knowledge is not necessarily deeply problematic—it is something that many scientists and philosophers are willing to acknowledge.[6]

Perhaps the most significant issue that arises out of thinking about scientific minimalism and quantum mechanics is the distinction as to whether science is or can pursue what is ontologically and metaphysically true versus pursuing what are effectively epistemological theories regarding what can be claimed with approximate or effective "models" of the world. It appears quantum mechanics can be interpreted as both—an epistemological and ontological theory. It seems both perspectives have something to offer, and scientific minimalism as defined here can help navigate this landscape.

---

[5]This quote is often attributed to Richard Feynman, but it appears it may have originated with David Mermin (2004). Stephen Hawking also takes issue with trying to more broadly "interpret" quantum mechanics and is quoted as saying: "When I hear of Schrodinger's Cat, I reach for my gun. All that one does, really, is to calculate conditional probabilities—in other words, the probability of A happening, given B. I think that that's all the many-worlds interpretation is. Some people overlay it with a lot of mysticism about the wave function splitting into different parts. But all that you're calculating is conditional probabilities" (Ferris 1997, p. 345).

[6]Stephen Hawking has noted, "I don't demand that a theory correspond to reality because I don't know what it is. Reality is not a quality you can test with litmus paper. All I'm concerned with is that the theory should predict the results of measurements. Quantum theory does this very successfully" (Hawking and Penrose 1996, p. 121).

The kind of scientific minimalism suggested here leans mostly toward the idea that even fundamental physics is more about epistemology than it is about ontology. Scientific minimalism is more about modeling the world's behavior, including observers in that world—perhaps, but not necessarily, leading to increasingly better approximations of an underlying reality (i.e., increasing verisimilitude). Because this definition of scientific minimalism includes the idea that reality can increasingly be more effectively modeled, a better understanding of reality can, in principle, be increasingly improved as more is learned about the behavior of the world. Because this version of scientific minimalism allows for the potential to increasingly approximate the truth, both possibilities seem plausible. Physical theories like quantum mechanics can be both epistemological and perhaps increasingly ontological as the world is modeled better over time—consistent with the idea of increasing verisimilitude as an outcome of science over the longer term. Scientific minimalism reflects a balance between humility in the face of evidence revealing a complex and counter-intuitive world and the increase of such evidence that may still nevertheless indeed tell us increasingly more about the physical world.

This is obviously just one assessment method for quantum theory in the context of scientific minimalism. One can of course choose many other evaluation criteria and/or assess the criteria differently. What is attempted here is only meant to be a rough assessment to help provide guidance and context for further evaluation of concepts related to quantum mechanics and cosmic evolution, particularly concepts that involve intelligent observers, some of which have already been touched on. We are hopefully now in a better position to make an informed assessment of provocative suggestions regarding how observers might affect the past and future evolution of the universe.

4.8 Participatory Observership and Cosmology: "Emergence" of Physical Laws?

John Wheeler reached back to (a) analysis by John von Neumann who developed the idea that the Schrödinger equation could be applied to the whole universe (von Neumann 1932) and to (b) subsequent developments from Eugene Wigner (Wigner and Margenau 1967) that emphasized how a human observer plays a critical role in wave function collapse. Wheeler extended this chain of "Copenhagen-like" thinking by suggesting that observers can influence the distant past, not just the immediate past associated with observer-dependent wave function collapse (Wheeler 1988).

4.8.1 Reaching “Inside” the Universe?

The observer-dependent retro-causation noted above, when applied to the universe as a whole, could possibly explain how the universe was brought into existence via a kind of participatory anthropic principle. As noted in Chap. 2, Paul Davies goes a substantial step further and speculates that the observer-dependent retro-causation of a “participatory anthropic principle” could possibly be applied to causing the physical constants and laws of physics themselves to emerge. Here, we see how earlier interpretations of quantum mechanics, particularly Copenhagen-like interpretations, extended by von Neumann, Wigner, and Wheeler, provide a conceptual foundation for some unusually provocative suggestions about the origin of the universe and possibly physical laws themselves (for example, from Paul Davies as noted in Chap. 2).

These kinds of suggestions can be interpreted as providing scientific support for the idea that mind and consciousness are fundamental properties of the world—similar to older philosophical ideas such as panpsychism. However, given the assessments provided earlier about the Copenhagen interpretation that question its validity, scientific minimalism would eschew substantial speculation that is built on what appears to be the insufficient foundation of Copenhagen-type interpretations of quantum theory. A participatory anthropic universe that has observers effectively “retrocausing” physical constants and laws of physics themselves is a significant stretch from what are arguably already problematic conceptual interpretations of quantum mechanics. Scientific minimalism would generally not embrace such extrapolations.

However, in the spirit of Occam’s razor, Davies does make an attempt to find explanations for the universe that do not apparently require anything “outside” the universe. His suggestion, however speculative by his own admission, at least conceptually allows for the possibility that the universe can in some sense explain itself via the emergence of sufficiently advanced observers that eventually exist at large enough scales to bring the entire universe into observer-participancy. This cosmic-scale observer-participancy can possibly then “retroactively” give rise to physical laws that allow such observers to emerge. But Occam’s razor can be interpreted in many ways. It does not necessarily have to be interpreted as only being concerned about the number of assumptions or whether causes are internal or external. It can also be interpreted to be concerned about the *kinds* of assumptions and implications as well. Observer-dependent retro-causation is an extraordinary mechanism based on what are arguably extraordinary assumptions that do not seem to have the commensurate extraordinary evidence to support them (at least not yet).

4.8.2 Applying Scientific Minimalism

Scientific minimalism would lean toward mechanisms that, on balance, appear to be less unusual than others. It may be that for now, all the concepts, mechanisms, and implications regarding fundamental physics, cosmic origins, etc., are all quite unusual (e.g., many-worlds, high dimensionality of string theory, etc.) and may be in some sense equally unusual. Scientific minimalism as defined here, in leaning toward functionalism, epistemological prudence, and increasing verisimilitude, would still allow for open-minded skepticism in considering provocative speculative ideas of the kinds explored here.

One way to facilitate that kind of open-minded skepticism for unusual ideas is to explore if and how ideas can be tested, or even more specifically, falsified (Popper 1965), however theoretical or physically difficult it may seem. Falsifiability, or "testability" more generally, is often seen as a hallmark of science, and scientific minimalism, requiring a high level of epistemological prudence, would want to seek at least a conceptual strategy for assessing the testability of an unusual idea. In the absence of epistemologically prudent test strategies, scientific minimalism would tend to be agnostic about very unusual conceptual ideas that appear to be untestable and even more suspicious about any deeper philosophical questions that highly unusual theoretical ideas may be associated with.

But how unusual is the idea of observer-related retrocausation? And are we really talking about causation "back in time"? The "observer" part of observer-related retrocausation is at least grounded in a history of seemingly plausible interpretations of quantum mechanics. However problematic that kind of interpretation may be, it is arguably not completely novel or completely unaccepted by scientists. What is more unusual is the extension of this kind of observer influence much further back in time—as Wheeler proposed. But how unusual is this idea? If we consider that time appears to be a variable in the universe, or at least a variable construct in some of our successful theories, and if we consider that laws of physics in general appear not to be strictly time-dependent, then it may not be so radical to see time, and hence causation, in different and highly counterintuitive ways. Time may not be relevant in the way we normally conceive of it, and laws of physics seem to allow for this theoretical possibility in general. Davies writes:

> When considering the universe as a whole, time loses its meaning, for there is nothing else relative to which the universe may be said to change. This "vanishing" of time becomes explicit when quantum mechanics is applied to the entire universe, and has been the subject of much discussion. To recover the notion of duration of time, one may consider the universe to be separated into two subsystems: an observer with a clock, and the remainder. The observer may then gauge the passage of time relative to the evolution of the rest of the universe by inspecting the clock, i.e., by making a quantum measurement on the clock variables that are correlated with some degrees of freedom of the rest of the universe (such as the size of the universe). Linde expresses the fundamental role played by the observer in recovering the passage of time from a "frozen" quantum universe: 'Thus we see that without introducing an observer, we have a dead universe, which does not evolve in time.... In the absence of observers, our universe is dead.' (Davies 2009, 386)

So, there does seem to be some plausible theoretical basis to think of time and causation very differently than we perceive it. But when Davies takes this large step further and suggests intelligent beings might retroactively influence or cause the emergence of specific laws and physical constants, he is proffering an idea that seems to lack enough evidence or suggestions for testability. As a result, epistemological prudence and functionalism of scientific minimalism would tend not to support it.

4.8.3 *Psychology of Science and Scientific Minimalism*

The considerations of epistemic prudence, and the often invoked "Occam's razor," can lead us to some psychological factors noted previously and may shed further light on these deliberations. One psychological factor has to do with the appeal of Occam's razor itself. The tendency to want simple, complete, "beautiful", understandable models of the world are arguably psychological predispositions that can vary across different personalities, even from one scientist to the next, and may have some basis in biological evolution.

In an attempt to limit an explanation about the existence and nature of the universe to not include anything outside the universe, Davies may force a path of inquiry that is more about satisfying a personal aesthetic than it is about remaining agnostic and objectively following where evidence or other forms of counterintuitive reasoning may lead. This is not unreasonable of course—it is a justifiable tendency that attempts to avoid positing unusual ontologies and "supernatural" forces or other extra parts of reality. Nevertheless, these kinds of leanings may be psychological factors to consider when thinking about unusual ideas and how scientific minimalism might view them. A psychological lens can inform concrete choices about what should be done about such ideas, including individual choices about areas to pursue, as well as how to handle funding and potential research programs that may involve substantial commitments and competition with other research needs.

Related to these kinds of aesthetic or psychological factors, we see what are arguably additional psychological leanings that may inform theory generation. Davies writes:

> Somehow, the universe has engineered not only its own self-awareness, but its own self-comprehension. It is hard to see this astonishing property of (at least some) living organisms as an accidental and incidental by-product of physics, a lucky fluke of biological evolution. Rather, the fact that mind is linked into the deep workings of the cosmos in this manner suggests that there is something truly fundamental and literally cosmic in the emergence of sentience. (Davies 2009, 385)

From a scientific minimalist standpoint, there appear to be notable leaps in this articulation. Davies says that it is "hard to see" how comprehensibility of the universe for beings like ourselves is a lucky fluke of evolution. But some amount of comprehensibility, or even high levels of comprehensibility, do not have to be a "lucky fluke" to be explained. It could conceivably be explained, at least partly, by biological evolution. It may in fact be the opposite of a lucky fluke. Brains, produced by the process of natural selection over a very long time in a dynamic environment like Earth's, could reasonably be expected to evolve capacities needed to "model" the world reasonably well—especially as we gain more time to carefully put our modeling minds to work on the world.

It is reasonable to think that exquisitely sensitive pattern recognition capabilities, along with flexible processing of possibility spaces associated with pattern recognition, can biologically evolve to help increase knowledge of the environment. Some ability to work with symbolic abstraction (Deacon 1997) is presumably evolutionarily useful and could be partially selected for since using symbols are an effective way to handle communications and interactions in increasingly larger and complex groups. Symbolic abstraction that can help explain and predict physical phenomena (e.g., mathematics) that are then tested with others for reliable results would not necessarily be so surprising in this kind of evolutionary context. If it is true that science can generally approximate the behavior of the world without necessarily ever discerning what is fundamentally true, then this would be consistent with the idea that intelligent creatures can create approximate models with abstract symbols that recognize useful patterns in the world without having the need to find the complete truth about the world. Indeed, evolution may not require objective truth for individuals and groups to be effective in surviving and reproducing.

In fact, as touched on previously, exquisitely sensitive pattern-seeking tendencies of human beings are famously capable of seeing patterns and "deeper truths" where there may be none—which can often nevertheless have fitness value to an individual or group despite the uncertainty or falsehood of a perception or belief. Extending this practical capability to even more abstract mathematical understandings of the world that may be less useful on a day-to-day evolutionary basis (e.g., theories of relativity and quantum mechanics) could then just be a result of applying basic capabilities that evolution has already equipped us with for fairly straightforward evolutionary reasons. In this sense, it does not seem that hard to see how we could have accumulated the understanding of the world that we have.

Davies also writes:

> The claim that life and mind are intrinsically fundamental to the workings of nature, and not irrelevant aberrations, implies that they are "written into" the laws of the universe and therefore are the expected, even inevitable, product of the outworking of these laws. This point of view is sometimes called the strong anthropic principle, and has received support from some prominent scientists. Thus Freeman Dyson has famously written: "As we look out into the universe and identify the many accidents of physics and astronomy that have worked together to our benefit, it almost seems as if the universe must in some sense have known we were coming." Likewise the Cambridge biologist Simon Conway Morris says that, "there is, if you like, seeded into the initiation of the universe itself the inevitability of intelligence." The strong anthropic principle also conforms with the widespread belief that the emergence of life is somehow inevitable because it is "built into" the laws of the universe. Christian de Duve calls life "a cosmic imperative." The biophysicist Stuart Kauffman echoes Freeman Dyson by declaring that we are "at home in the universe." (Davies 2009, 386)

Again, scientific minimalism would not suggest these are necessarily unreasonable or dismissible speculations per se, but epistemic prudence and functionalism would be very sensitive to strong claims that are arguably more philosophical than scientific. As touched on earlier, epistemic prudence might also prompt questions about psychological predispositions that might lead to advocating physical worldviews that place life and intelligence in privileged positions.

Despite this skeptical application of scientific minimalism, Davies' reasoning that a universe may bring itself, including its laws of physics, into existence through a kind of retroactive observer influence, seems at least noteworthy given a history of ideas that can be invoked to support it—even as a remotely plausible theoretical possibility.

However, while it seems plausible enough not to dismiss, a scientific minimalist worldview would tend to see such an idea as going further than is warranted.

4.9 Psychology of Science

Taking the previous discussion further, we can more broadly consider potential psychological factors that are involved in scientific pursuits. An interesting thread in the discussion of scientific minimalism and its relationship to epistemology and ontology is how much of it may be wrapped up in human psychology—particularly evolutionary psychology and social psychology. The "psychology of science" relates to the philosophy of science and can further inform how we assess broader implications of quantum theory, scientific minimalism, and worldviews more generally.

The particular positions taken by Bohr and Einstein, the varying levels of conviction they and others have had, and the subtle and varying degrees of bias that may be involved, could have strong links to basic individual human psychology and may not be controllable or even noticeable to individuals or practicing scientific communities as much as we might think—despite training and other forces that are intended to balance against personal and collective bias (Kelemen et al. 2013). Here, we will briefly note a few potential psychological factors that may be worth keeping in mind as we go forward to contemplate broader considerations in evaluating science, values, and intelligence in the context of cosmic evolution.

Predispositions for accepting uncertainty about knowledge—not just about the fundamental workings of the world, but about anything in human experience—presumably play a role in how scientists view and practice the scientific pursuit. Trying to explain all of reality may be a noble pursuit by many standards, but it can also be problematic and potentially dangerous depending on the degree of dogma and idealism involved—and depending on how scientific knowledge is used. Modern science has a fairly good track record overall, but certain kinds of speculation, even today when there are reasonably good checks and balances, can create false impressions and provide unstable foundations for problematic beliefs.

At the heart of much human psychology is individual ego, driven largely by biological self-interest. While science is meant primarily as an attempt to pursue an objective self-correcting process, there is reason to think it often falls short of that standard, including complex and subtle social factors, however self-correcting it may be in the long-run (Kuhn 1962; Collins 1983; Kitcher 1993). Some of the shortcomings can arguably be tied to human ego and the psychological (and evolutionary) importance of being a credible source of information in a complex world and complex social settings. Being a "good" source of information can increase social status—an important evolutionary dynamic and selection pressure that can enhance individual and group fitness (Cheng et al. 2014).

Sometimes, ideas become an important part of a person's identity, particularly if a person creates the idea and particularly if that idea serves other purposes that serve self or group interests. Defending identity, even at high cost, is arguably a helpful evolutionary strategy and hence an important human psychological need. Ideas may come often merely in the service of, and in the defense of, personal identity.

Related to the need for defending individual identity is the need to defend group identity. Bound up in group identity (as well as individual identity) are often perceptions of privileged positions of humans in a worldview. While there has been a history of so-called great demotions that science has delivered to humanity (Sagan 1994), there is nevertheless still great appeal in placing human beings (including human consciousness) in an important role within a worldview. That appeal may or may not lead to reasonable or true conceptualizations of broader worldviews, but the compulsion is arguably still deeply embedded in human psychology—again, perhaps for evolutionary reasons. For example, it is reasonable to consider that conceptions of "observer-dependence" that give rise to essentially "co-creating" reality are appealing in that they place human beings in an elevated, if not critical, position in a worldview.

Another factor that directly relates to the previous discussion and to some of the assessments of the more extreme versions of observer-dependence is the allure of novel, mysterious, and unusual "revolutionary" ideas. Some may even see certain ideas as "romantic" in the broader inspirational sense. Revolutionary ideas are stimulating, intellectually fun, and entertaining—particularly when captured well in works of fiction. The theory of special relativity was a compelling example of how unusual, counterintuitive, and complex scientific ideas can be. It, and quantum theory, may have fully swung open a door that was already psychologically ajar, allowing the modern scientific mind to entertain just about any possibility.

Bohr and others of the Copenhagen interpretation may ultimately be correct about the fundamental counterintuitive nature of the world and the nature of the scientific pursuit. It may be that Bohr and others made very thoughtful objective assessments based on what they knew and what the evidence was suggesting, but it does appear that unwarranted conclusions were reached and strongly defended. This could be a result of psychological considerations such as wanting to identify with a truly revolutionary idea or dramatic moment in human history. Or maybe, it is partly explained by having a sufficiently open-minded psychological predisposition that better tolerates extreme counterintuitive evidence and uncertainty. Such an orientation may also be more receptive to a purely functional approach to science, (as many scientists seem to reflect).

Simplicity and completeness are also appealing to human minds for psychological reasons, and the invocation of epistemic prudence in evaluating interpretations of quantum mechanics may indeed be an example of using a misguided "metric" driven by psychological predispositions for relative simplicity. It may be that the universe, or whatever ultimate reality exists, is indeed consistent with seeking simplicity and completeness, but by many standards, much science appears to be complex and counterintuitive. It may be the case that complete and fundamentally true scientific views of reality are at least theoretically within reach, but some interpretations of science can also reasonably suggest that such pursuits, while worthwhile and noble, may ultimately be futile.

As so much science has taught us, particularly modern science, physical reality very often defies human preconceptions, many of which are rooted in human psychology. Ultimate reality eludes much human intuition and often presents us with what are in retrospect, illusions. The history of science has indeed sensitized us to the folly of human bias and how the universe can constantly surprise us.

4.10 Summary of Chapter 4: Scientific Minimalism

4.10.1 Science Informs Values, But…

Scientific minimalism would generally not claim much regarding human values and any normative aspiration of human beings. Stronger versions of scientific minimalism could however suggest that science can, and perhaps to some extent *should*, help inform what values might be best for human beings and how we might go about pursuing those values—e.g., at least partially, but not exclusively informed by evolutionary psychology. Scientific minimalism would tend to encourage a high degree of caution regarding the appeal of certain philosophical ideas and values and our pursuits to find ultimate patterns, meaning, and purpose. We also briefly touched on how values, or psychology more generally, can have undue influence on science, such as defending individual and group identity, including defending privileged roles for groups and human agency more broadly.

4.10.2 Quantum Theory

Scientific minimalism would suggest that quantum theory as we understand it today is almost certainly incomplete and more unified theories should be pursued. However, scientific minimalism also suggests that too much speculation or concern about the broader conceptual or intuitive implications of various interpretations of quantum theory may not be as important as some think—and may even be misguided. Epistemological prudence and functionalism are important and can be leveraged in the pursuit of increasing verisimilitude despite significant challenges (and perhaps fundamental limitations) of knowing the complete truth about the world. Our foray into the Copenhagen interpretation of quantum theory suggests that scientific minimalism is at least agnostic about its interpretive utility, if not dismissive of its broader philosophical claims.

4.10.3 Cosmology and the Participatory Universe

Any undermining of the Copenhagen interpretation also then calls into question much broader stronger cosmological claims that are derived ultimately from it. For example, suggestions that intelligent observers play a critical role in bringing the universe into existence, and also potentially bringing physical laws into being, would be considered highly suspect. Scientific minimalism would however recognize a critical causal relationship of life and human intelligence to cosmic evolution as products of cosmic evolution, but would generally not imbue that relationship with any particular value per se, other than to recognize whatever minimal value there may be associated with the physical causes of our existence. That is, our causal emergence from cosmic evolution is notable and important for our understanding and possibly important in recognizing a critical relationship between ourselves and the universe (as indicated in Fig. 1.1 by the overlap of scientific minimalism with the relationalist framework that will be discussed in the next chapter), but it is not necessarily important in terms of human motivation or values. Scientific minimalism, which could also be associated with a general form of philosophical minimalism, is essentially agnostic about a larger and/or specific "objective" meaning or purpose for the universe and intelligence.

4.10.4 Relationality

The scientific minimalist perspectives of value, quantum theory, and cosmic evolution do seem to strongly suggest some utility for thinking in terms of relationships and "relationality" more generally as a kind of metaview. Indeed, we saw that science can relate to value by informing us about facts of the world (e.g., evolutionary psychology) and also by noting how at least instrumental value may ultimately be relational in the sense that value for something is usually realized relative to something else. We also saw that the relational interpretation of quantum mechanics appears to have substantial merit. Quantum entanglement may suggest a deep and fundamental sense of connection and relationality in the physical world more generally and will be touched on further in subsequent chapters.

As noted in Fig. 1.1, some interpretations of scientific minimalism, recognizing the dynamic expanding universe predicted by general relativity, can be interpreted to suggest forms of what might be called "weak relationalism" in the sense that we and all objects in the universe appear to have been created by an unfolding emergence of cosmic dynamics and relationships. We are importantly related not only to some "underlying" fundamental physical reality of the cosmos, but we are critically related to and dependent on the dynamics of that reality and its overall evolution. Scientific minimalism is also arguably consistent with the importance of seeing the world in terms of *process*, at least in the form of physical dynamics that unfold in the universe over time, without necessarily having to commit to strong specific claims about fundamental, ultimate, and absolute "material" objects in the world—something we will explore more in the next chapter on relationalism.

References

Albert, D. (1992). *Quantum mechanics and experience*. Cambridge: Harvard University Press.

Ayer, A. J. (1936). *Language, truth, and logic*. London: Victor Gollancz Ltd.

Ballentine, L. E. (1970). The statistical interpretation of quantum mechanics. *Reviews of Modern Physics, 42*(4), 358–381.

Bub, J. (1974). *The interpretation of quantum mechanics*. Boston: D. Reidel Publishing Company.

Cheng, J. T., Anderson, C., & Tracy, J. L. (2014). *The psychology of social status*. New York: Springer.

Collins, H. (1983). An empirical relativist programme in the sociology of scientific knowledge. In *Science observed: Perspectives on the social study of science* (pp. 115–140). London: Sage.

Davies, P. (2009). Life, mind, and culture as fundamental properties of the universe. In S. J. Dick & M. Lupisella (Eds.),*Cosmos and culture: Cultural evolution in a cosmic context*. Washington, DC: NASA History Series. https://history.nasa.gov/SP-4802.pdf.

Deacon, T. (1997). *The symbolic species: The co–evolution of language and the brain*. New York: W. W. Norton.

Dennett, D. (2001). The Zombie Hunch: Extinction of an intuition? In A. O-Hear (Ed.), *Philosophy in the New Millennium*. Cambridge: Cambridge University Press.

Denning, K. (2009). Social evolution. In S. J. Dick, & M. Lupisella (Eds.), *Cosmos and culture: Cultural evolution in a cosmic context*. Washington, DC: NASA History Series.

Deutsch, D. (1997). *The fabric of reality: The science of parallel universes and its implications*. New York: Penguin Books.

Doppelt. G. (1986). Relativism and the reticulational model of scientific rationality. *Synthese, 69*(2), 225–252 (Springer).

Durr, D., Zanghi, N., & Goldstein. S. (1996). Bohmian mechanics as the foundation of quantum mechanics. In J. T. Cushing, A. Fine, & S. Goldstein (Eds.), *Boston Studies in the Philosophy of Science.Bohmian mechanics and quantum theory: An appraisal* (Vol. 184, pp. 21–44). Kluwer

Einstein, A., Podolsky, B., & Rosen, N. (1935). Can quantum-mechanical description of physical reality be considered complete? *Physical Review, 47*(10), 777–780.

Einstein, A. (1949). Remarks concerning the essays brought together in this co-operative volume. In P. A. Schilpp (Ed.), *Albert Einstein: Philosopher-scientist*. New York: Harper & Row. 8th printing in 2000.

Everett, H. (1957). Relative state formulation of quantum mechanics. *Reviews of Modern Physics, 29*, 454–462.

Ferris, T. (1997). *The whole shebang: A state-of-the-univers(s) report*. Simon and Schuster.

Friedman, M. (1999). *Reconsidering logical positivism*. New York: Cambridge University Press.

Ghirardi, G. C., Rimini, A., & Weber, T. (1985). A model for a unified quantum description of macroscopic and microscopic systems. In L. Accardi et al. (Eds.), *Quantum probability and applications*. Berlin: Springer.

Ghirardi, G. C., Rimini, A., & Weber, T. (1986). Unified dynamics for microscopic and macroscopic systems. *Physical Review D, 34*, 470.

Gibbins, D. (1987). *Particles and paradoxes*. Cambridge: Cambridge University Press.

Harris, S. (2010). *The moral landscape: How science can determine human values*. New York: Free Press.

Hanfling, O. (2003). *Logical positivism* (p. 193f). Routledge.: Routledge History of Philosophy.

Hartle, J. (1968). Quantum mechanics of individual systems. *American Journal of Physics, 36*(8), 704–712.

Hawking, S. (1993).Is the end in sight for theoretical physics? *Black holes and baby universes and other essays*. New York: Bantam.
Hawking, S., & Penrose, R. (1996). *The nature of space and time*. Princeton University Press.
Healey, R. (1989). *The philosophy of quantum mechanics*. Cambridge: Cambridge University Press.
Jammer, M. (1974). *The philosophy of quantum mechanics*. New York: Wiley.
Kelemen, D., Rottman, J., & Seston, R. (2013). Professional physical scientists display tenacious teleological tendencies: Purpose-based reasoning as a cognitive default. *Journal of Experimental Psychology: General, 142*(4), 1074–1083.
Kitcher, P. (1993). *The advancement of science: Science without legend, objectivity without illusions*. New York: Oxford University Press.
Kuhn, T. S. (1962). *The structure of scientific revolutions* (2nd ed.). Chicago: University of Chicago Press (1970).
Laudan, L. (1977). *Progress and its problems*. Berkeley: University of California Press.
Laudan, L. (1984). *Science and values: The aims of science and their role in scientific debate*. Berkeley: University of California Press.
Maudlin, T. (1995). Why Bohm's theory solves the measurement problem. *Philosophy of Science, 62,* 479–483.
Mermin, D. (1991). Is the moon really there when nobody looks? Reality and the quantum theory. In R. Boyd (Ed.), *The philosophy of science*. Cambridge: The MIT Press.
Mermin, D. N. (2004). Could Feynman have said this? *Physics Today, 57*(5), 10–11.
Newton-Smith, W. H. (1981). *The rationality of science*. Boston: Routledge & Kegan Paul.
Omnes, R. (1994). *The interpretation of quantum mechanics*. Princeton: Princeton University Press.
Penrose, R. (2014). On the gravitization of quantum mechanics 1: Quantum state reduction. *Foundations of Physics, 44,* 557–575.
Popper, K. (1965). *The logic of scientific discovery*. New York: Harper Torchbooks.
Rovelli, C. (1996). Relational quantum mechanics. *International Journal of Theoretical Physics, 35,* 1637–1678.
Sagan, C. (1994). *Pale blue dot: A vision of the human future in space*. New York: Ballantine Books.
Saunders, S., Barret, J., Kent, A., & Wallace, D. (Eds.). (2010). *Many worlds? Everett, quantum theory, and reality*. Oxford University Press.
Schlosshauer, M., Kofler, J., Zeilinger, A., (2013) A snapshot of foundational attitudes toward quantum mechanics. *Studies in History and Philosophy of Science Part B: Studies in History and Philosophy of Modern Physics, 44* (3), 222–230.
Shermer, M. (2004). *The science of good and evil*. New York: Times Books/Henry Holt & Company.
Suppe, F. (1989). *The semantic conception of theories and scientific realism*. Chicago: University of Illinois Press.
Von Neumann, J. (1932). *The mathematical foundations of quantum mechanics* (R. T. Beyer, Trans.). Princeton Univ. Press (1996 edition).
Wallace, D. (2012). *The emergent universe: Quantum theory according to the Everett interpretation*. Oxford University Press.
Wheeler, J. (1988). World as a system self-synthesized by quantum networking. *IBM Journal of Research and Development, 32*(1).
Wigner, E., & Margenau, H. (1967). Remarks on the mind body question, in symmetries and reflections, scientific essays. *American Journal of Physics, 35*(12), 1169–1170.
Worrall, J.(1989) Structural Realism: The Best of Both Worlds?. *Dialectica, 43* (1–2):99–124. Reprinted in D. Papineau (ed.), The Philosophy of Science, Oxford: Oxford University Press, pp. 139–165.

Chapter 5
A Relationalist Framework

5.1 Background and Overview

In Chap. 3, we briefly explored value theory and touched on potential relationships between facts and values and the relational nature of instrumental value and of value more broadly.[1] The relational nature of value will be explored in general terms in this chapter and with further details as it relates to the three specific cosmological theories of value in subsequent chapters.

As suggested at the end of the previous chapter, scientific minimalism can be interpreted to have important "relational" characteristics or implications (suggested by its overlap with the relationalist framework in Fig. 1.1). Our foray into quantum mechanics suggests that something having to do with relations, or perhaps "relationality" more generally, seems to be at least epistemologically and scientifically important. In our upcoming discussion on relativity, we will see something similar. At the very least, it appears that some form of relational epistemology is warranted based on relativity and quantum mechanics. For example, the relation between "subject" and "object" has become increasingly important and perhaps increasingly merged into more systemic understandings of science and philosophy.

Here, we will develop a broad-based pluralistic and multifaceted relationalist framework (as shown in Fig. 1.1) that might also be thought of as a relational worldview of sorts. It will have a number of features and increasing degrees of admittedly speculative suggestions with potentially wide-ranging implications. This relational framework can provide a systematic conceptual foundation for addressing a number of interesting philosophical and scientific challenges.

---

[1]Moore (2004) explores these issues explicitly in the context of relationalism.

M. Lupisella, *Cosmological Theories of Value*, Space and Society,
https://doi.org/10.1007/978-3-030-25339-4_5

There is a long history of what can be referred to as "relational philosophy" or, in the indomitable spirit of "ism-izing," what could be called *relationalism* in the broadest sense.[2] Going back to the ancient Greeks and forms of Eastern thought such as Buddhism, there is a tradition of thought that emphasizes relations. As a starting point, relational thinking can be thought of as emphasizing the critical importance of relations in knowledge and reality, and in our experiences in the broadest sense. This would generally involve an emphasis on relations as opposed to conceptualizations such as "substantivism" which posits substantive "things" or objects (e.g., atoms) as a fundamental basis of reality. So in this context, relational theory is not just about discerning and studying specific relationships per se—as is done in so many domains of knowledge. Nor is relationalism simply about emphasizing the general observation that "everything is related." Relationalism is instead meant here to be a broad fundamental conceptual umbrella that emphasizes a critical role for *relations in particular* and *relationality in general* as an important, perhaps central and fundamental, feature of the world.

For some, relationalism might be thought of as a form of metaphysics—and we will explore that perspective more explicitly later. But even the term "metaphysics," as it has often been traditionally used in philosophy, does not necessarily sufficiently capture relationalism as it is being used in this broad context. As we will see, relational thinking can be extended to many spheres of experience and domains of interest. This may include more intangible and/or emergent elements of human psychology and other areas that have arguably not been fully treated in traditional philosophical metaphysics.

In the broadest sense then, we can think of relationalism as covering (a) relational epistemology, (b) relational metaphysics/ontology (e.g., Young 2011, Marmodoro and Yates 2016) and (c) "relational experience" construed broadly, covering aspects of life such as relations with others (e.g., Gergen 2009; Chen et al. 2011). Taken together, these areas can be thought of as constituting a relational framework or perhaps even a form of "relational worldview" (Spretnak 2011; Burkhart 2004).[3] Here, we will focus primarily, although not exclusively, on a and b above, which may have overlap with and implications for relational experience more generally.

To help get a sense of the landscape that involves an emphasis on relations, we will first touch on some relational thinking by briefly covering a few examples of relational thought, with an emphasis on process philosophy. We will then more deeply explore relational considerations with respect to the modern sciences of relativity and quantum mechanics, pointing to what appears to be a strong form of relational epistemology. Drawing partly from concepts of relational epistemology and other areas, we will then explore in subsequent chapters the potential for stronger relational claims, with an emphasis on a theoretical construct (the *connection-action*

---

[2]"Ontic relationalism" is a relational theory of being developed by Kaipayil (2009) but is a narrower use than intended here.

[3]Spretnak (2011) gives us a wide-ranging treatment of "dynamic interrelatedness" in her book, *Relational Reality: New Discoveries of Interrelatedness that are Transforming the Modern World*. Brian Yazzie Burkhart (2004) tells us that much Native American philosophy often centers on a "principle of relatedness."

principle) that suggests how a relational model of reality may have merit and how that relationality might be manifested as a dynamic creative universe.

5.2 Process Philosophy

> The Universe is made of stories, not of atoms.
>
> —Muriel Rukeyser (1968)
>
> What has been lost sight of is that physics as a subject of thought is a dynamic interplay between storytelling and equation writing. Neither one stands alone, not even at the end of the day.
>
> —Christopher Fuchs[4]
>
> At the end of the day, or at the end of your life, it doesn't matter much that you were happy much of the time. Wouldn't you rather have a good story to tell?
>
> —Sean Carroll (2016)

Perhaps the most well-known philosophically rigorous form of relational thinking is "process philosophy" which arguably has its first articulations in both Eastern and Western thought going back to Buddhism and Daoism in the East and to the ancient Greeks in the West. More explicit and systematic modern treatments in the West can be traced to Georg Hegel, Charles Pierce, William James, Henri Bergson, and John Dewey (Rescher 1996), with perhaps the most systemic and influential articulation coming from Alfred North Whitehead and his "philosophy of organism." While Whitehead himself appears to have never used the phrase "process philosophy," he did invoke a "principle of process" when he wrote of an entity that "Its 'being' is constituted by its 'becoming.' This is the 'principle of process'" (Whitehead 1929, 23).[5] In this way, Whitehead does not prioritize being over becoming (Kraus 1998).

In its simplest form, process philosophy emphasizes the *dynamic* nature of reality. As indicated above by the quote from Whitehead, process philosophy emphasizes "being as becoming" through an endless unfolding set of related dynamic processes. Of particular note for the purposes of this book is Whitehead's emphasis on "creativity" as the ultimate reality, the ultimate source of all that is. Indeed, it appears that Whitehead coined the term "creativity" (Meyer 2005). Creativity is the most general and most fundamental source and category of reality from which everything flows, even "God." Novelty is a key manifestation of creativity for Whitehead. Process philosophy has given rise to many wide-ranging articulations (e.g., Lucas 1989; Mesle 2008).

Whitehead's process philosophy is often bound up in his and others' conceptions of God, and to this day process philosophy has a strong following among theologians and has given rise to the broad field of process theology (e.g., Hartshorne 1971),

---

[4]Quote from Fuchs interview in Schlosshauer (2011a, b, p. 71), *Elegance and Enigma: The Quantum Interviews*.

[5]Notably, the subtitle to Whitehead's (1929) book, *Process and Reality*, is "*An Essay in Cosmology*."

including intersections of feminist thought and theology (Ingram 2006). However, there are also treatments of process philosophy that do not emphasize theological considerations, and they often include key distinctions between teleological views (e.g., which can include theological views) and more "secular" or naturalistic views that suggest a lack of directionality (Rescher 1996). In subsequent chapters, we will explore views that lie somewhere between these views.

As a hint of things to come, we can note this from Sean Carroll:

> The mistake we make in putting emphasis on happiness is to forget that life is a *process*, defined by *activity and motion* (italics added), and to search instead for one perfect state of being. There can be no such state, since change is the essence of life. (Carroll 2016, 426)

Indeed, while it is a notable leap, change may be the essence not just of life, but of the universe itself—for which activity and motion play key roles—as will be explored further in Chap. 8 on the connection-action principle.

5.3 The Relational Revolutions of Relativity and Quantum Mechanics

Given the unusual and seemingly revolutionary nature of relativity and quantum mechanics, and given that this section on developing a relationalist framework relies heavily on those scientific framworks, we will spend a bit of time assessing their revolutionary nature. We will explore how taken together these modern scientific enterprises can be seen to represent a fundamental shift in thought to a kind of "observer-relation" revolution, providing support for the importance of a relational framework more generally.

5.3.1 Scientific Revolutions

Philosophy and the history of science have provided many characterizations of scientific revolutions. Three articulations will be drawn from here: (1) the research program of Imre Lakatos, (2) the problem-solving efficacy of Larry Laudan, and (3) Philip Kitcher's consensus practice. These three accounts have some similarities and are essentially rationalist. Each account is complex and involves understanding science in a comprehensive way; however, for the purposes of this treatment, we will only be concerned with appealing to what is needed to construct a very basic view of scientific revolutions to the extent that they can then be distinguished from, and inform, philosophical, epistemological, and metaphysical revolutions.

Modifications of theories in the face of anomalies or other challenges give rise to what Lakatos (1978) calls a scientific research program (SRP) which consist of a series of related theories. Theories make up SRPs, and SRPs are the units to be evaluated when comparing and ultimately choosing between sets of scientific ideas.

SRPs have three primary elements: (1) a hard core, (2) the negative heuristic, and (3) the positive heuristic. The hard core is what matters most for the purposes of this treatment. It consists of theoretical assertions which all theories in the SRP share. Indeed, it is so "hard" that the SRP appeals to the negative heuristic as the mechanism by which it is preserved in the face of anomaly. Similarly, the positive heuristic protects the hard core by providing some guidelines on how to adjust the auxiliary belt of hypotheses (which "surround" the hard core) in order to accommodate a given anomaly. A good example of a hard core is the universal law of gravitation and the three laws of dynamics in Newtonian mechanics (Lakatos 1978). It is not clear whether Lakatos would include certain metaphysical notions as part of a hard core. A simplified, but valid characterization for scientific revolution in Lakatos' view amounts to equating a scientific revolution with a transition from one SRP to another.

As touched on previously, Larry Laudan sees science as primarily a problem-solving endeavor, meant to constantly increase the problem-solving efficacy of research traditions (arguably consistent with increasing verisimilitude explored previously). Problems are normally defined by their perception and formulation by the scientific community for which he claims there is generally a consensus. Additionally, problems are generally rooted in an empirical foundation. Laudan says, "Empirical problems are first order problems: they are substantial questions about the objects which constitute the domain of any given science" (Laudan 1977, 15). Scientific revolutions for Laudan generally amount to a shift to a whole new, or at least substantially different, set of problems. Newton-Smith sees the empirical basis as crucial and notes that Laudan's use of "empirical" is intended to distinguish between empirical and "such conceptual problems, as say, the questions of the intelligibility of absolute space" (Newton-Smith 1981, 191). This latter observation is important because it helps us distinguish between scientific and philosophical revolutions.

Kitcher (1993) emphasizes the behavior of the community as opposed to the individual, which he believes are often inappropriately conflated. Kitcher's consensus practice consists of: (1) a language, (2) an impersonal assessment of significant questions, (3) a set of accepted statements with a partial justificatory structure, (4) a set of explanatory schemata, (5) a set of paradigms of authority and criteria for identifying authorities, (6) a set of exemplary experiments, observations, and instruments and justificatory criteria, and (7) a set of methodological exemplars and methodological principles. Item 2 is essentially Laudan's problem solving extrapolated to the consensus of the community. Item 3 is similar to Lakatos' hard core. Item 4, and possibly item 5, can be interpreted as Lakatos' belt of auxiliary hypotheses and positive heuristic. Items 6 and 7, a set of exemplars, are most important for the purposes of this treatment. It should be noted that while they are similar to Kuhn's paradigm, they offer no commitment to the epistemological differences resulting from Kuhnian attempts to divide the progression of scientific theories into distinct and separate, often incommensurable, episodes (Kuhn 1962). A scientific revolution on this view involves the complete replacement of one or more of the above elements.

From these three views, we can construct a view of scientific revolutions that will help us distinguish them from, as well as inform, philosophical revolutions.

Lakatos gives us the scientific research program which has, as its heart, a hard core of formal, generally theoretical postulates, which, in the course of normal science, the wider scientific community sees as unassailable. For our purposes, the hard-core element of the SRP captures the working theoretical principles (mathematical or otherwise) to which scientists generally constrain themselves, intentionally or not. "Working" is an important word here because this core must have direct operational significance for the scientist (consistent with functionalism of scientific minimalism noted previously). That is, it must have direct applicability and translatability into empirically based efforts and daily activities such as observation, experimentation, and data analysis. A significant modification or replacement of the hard core is one major element of a scientific revolution that this hybrid view advocates.

This leads us to two other important aspects of scientific revolutions contributed by problem-solving and consensus practice. The set of problems (which are generally empirically rooted) that are deemed important must be supplanted by new problems, either because the old problems have become irrelevant or because they have been answered. Similarly, this hybrid view of scientific revolutions requires the consensus practice of a community to also be supplanted for the same reasons noted for problem solving—with the primary emphasis on the set of exemplary experiments, observations, and instruments and justificatory criteria and the set of methodological exemplars and methodological principles. Scientific revolutions, in this view, are significant shifts or replacements of the scientific community's working-level activities: (a) their theoretical working principles which have a direct impact on empirical efforts, namely (b) the problems to be considered and (c) the practices adopted by the wider community.

5.3.2 *Philosophical Revolutions*

One important aspect of Thomas Kuhn's conception of scientific revolutions is the significant shift or complete change in worldview, which often results in "incommensurability" (Kuhn 1962). However, such shifts in worldviews, if we understand worldviews to be contextual, or metaphysical, as Kuhn indicates, seem to transcend the relevant scientific elements for any given revolution. For example, changes in the perceptions of God or humanity's place in the universe that may have resulted from shifting to a heliocentric model are not, strictly speaking, just scientific changes. They are changes in worldview.

Changes in worldview, e.g., suggestions of an epistemological or metaphysical nature such as Ockham's razor or the theory of minima naturalia (see Dijksterhuis 1950), have had an important role in giving birth to many scientific theories. Similarly, corpuscularism, reaching its height with Cartesian physics, was essentially a metaphysical view that had tremendous impact on many areas of science (McAllister 1996). To help us establish a working characterization of philosophical revolutions, we will look briefly at (1) Kuhn's conception of worldview shifts, (2) some of Paul

Thagard's key elements of conceptual revolutions, and (3) James McAllister's role for aesthetics in scientific revolutions.

Kuhn's scientific revolutions involve "changing the world"—or at least changing worldviews to the extent that those changes cannot be directly compared with rival worldviews, since there would be little or no commonality between them—i.e., they would be incommensurable. The concept of revolutions suggested in this treatment can be characterized by shifts in worldview, but not necessarily (or even generally) by the incommensurability claimed by Kuhn. So what is a worldview? One way, among others, that Kuhn talks of a worldview is that which characterizes humanity's relationship to the universe and to God (Kuhn 1957). More generally, we might see a worldview as that which defines a person relative to a larger context, whether that be the universe or some larger domain including the natural world, God, a Platonic realm, or "spirit" of some kind. At the very least, we can take "worldview" to mean what it sounds like, a "view of the world," where "view" implies a particular view from "somewhere" providing a particular perspective, and "world" implies basically everything. So "worldview" can be seen as a fundamental, coherent, or "complete" view of everything that matters to a person or group (Lupisella 2019).

In *Conceptual Revolutions* (1992), Thagard (1992) defines conceptual systems as complex computational structures and hierarchies of parts. Revolutions often involve tree jumping which is the shifting of a concept from one branch of a tree to another and tree switching which is changing the organizing principle of the hierarchical tree. The latter is most relevant for defining metaphysical revolutions. Thagard sees tree switching as the most dramatic change to a conceptual system and as such is most closely aligned with the preferred conception of a metaphysical revolution being developed here. Tree switching most often results in a radically different hierarchy of conceptual organization from that which it replaced.

James McAllister sees scientific revolutions as entailing "aesthetic ruptures" or the abandonment of aesthetic principles. Such aesthetic principles can vary widely, but some examples are symmetry, simplicity (e.g., Occam's razor as noted previously), and religious views. As a revolution mounts, previously applied aesthetic principles and constraints are repudiated. In fact, the break comes from a battle—not necessarily between two competing aesthetic principles, but between those who choose the empirical evidence and those who choose the preservation of an aesthetic principle.

From the above considerations, we can suggest that philosophical revolutions at least involve significant changes or replacements of (1) worldviews, (2) conceptual systems, and (3) aesthetic principles. These general categories can be broken up further to capture more specific manifestations of philosophical revolutions. Under worldviews, we might include religious views, epistemological orientations, and conceptions of value. Indeed, Kuhn specifically notes how the Copernican Revolution was also a part of a transition in our sense of values (Kuhn 1957). Under conceptual systems, we might include ontological claims, knowledge-organizing principles, and language rules. Aesthetic principles might include simplicity, symmetry, and something as general as a conservative orientation that tends to preserve status quo until it is absolutely necessary to abandon it—i.e., do not fix what is not completely broken.

The broad nature of philosophical revolutions prompts us to consider a fourth criterion for defining such revolutions; namely, the knowledge that philosophical revolutions reveal should be applicable and presumably influential in other areas outside certain scientific disciplines and outside of science more generally. The reverse is also true: philosophical revolutions stemming from other areas of knowledge can affect science as well.

5.3.3 The Relativity Revolution

Einstein saw his theory of relativity as a natural extension of Newtonian mechanics (Thagard 1992; Cohen 1985). However, there are significant differences between Newtonian mechanics and relativity that are worth noting to help support the broader relationalist framework being explored here. Einstein's theory of relativity involves significant shifts in all three areas of scientific revolution discussed earlier. In considering the shift to relativity theory, we should think of pre-relativity physics as both Newtonian mechanics and Maxwell's electrodynamics. In this light (yes, "pun" intended), pre-relativity physics can be said to have contained (a) Newton's three laws of motion and his universal law of gravitation (Lakatos 1978), (b) absolute space and time, (c) Maxwell's equations of electrodynamics, and (d) the luminiferous aether by which light was thought to have propagated and was also thought to have represented an absolute frame of reference (Thagard 1992).

Relativity theory resulted in significant changes, if not the complete replacement of Newton's laws of motion and gravity as well as the abandonment of the luminiferous aether and absolute space and time. All of Newton's laws were replaced by Einstein's formulations, and the aether was shown to have been superfluous. These significant changes in theoretical working principles then affected the practices of scientists for the decades that followed (Stachel 1990), hence demonstrating the significance of the shift.

Stanley Goldberg's distinction between Newtonian mechanics and relativity suggests that Newton's theory was an ontological theory, while Einstein's was theory of measurement (Goldberg 1984). Einstein was in some sense forced to adopt a theory of measurement in order to explain the picture he saw emerging from the state of physics at the time. One key mechanism by which Einstein arrived at his conceptions is his extensive consideration of simultaneity, or more fundamentally, the limitations of information communication via the finite speed of light. Einstein realized that information could not be communicated any faster than this. This realization played a critical role in the construction of the special theory of relativity and resulted in a theory that was fundamentally about how the world is measured—i.e., about how information about the world is communicated to an observer.

The shift from an ontological theory to a theory of measurement reveals much for the kind of change that transpired. It was a shift which suggested that physical theories might be better approached as theories about how we measure the world, or how we *know* the world, as opposed to theories about the fundamental nature of the universe. In this way, the relativity revolution was fundamentally a shift in how physical theories are constructed and evaluated. That is, it appears to have been at least a dramatic *epistemological* shift, if not more. This kind of shift counts as an important philosophical revolution derived from a scientific revolution.[6] Inextricably bound up with this epistemological shift is the important role of the observer, since it is the observer which is a critical component of measurement. This is realized by noting an important detail of relativity, namely, that the actual values of measurements depend on the frame of reference of the observer. This suggests at least an "observer-relation" aspect of the theory, or possibly a form of "observer-dependence," depending on how far one wants to go in describing the role of the observer.

We should further note that Einstein's general relativity broke dramatically with previous conceptions of space and time by showing how they were both intimately *related* to each other. Einstein famously combined space and time into a ubiquitous geometric construct, "space-time," and showed how gravity can be seen as a kind of warping of space-time. Einstein's insight was to leverage the possibilities for how space, time, and gravity were related to each other, and those relationships are key today for understanding, predicting, and achieving complex practical pursuits such as space missions and detecting black holes. As special and general relativity shows, much of the history of science has been to reveal how apparently disparate details and areas of knowledge can be *related* to each other and, in some cases, be formally united via those relationships for more comprehensive powerful theories of the world.

5.3.4 The Quantum Revolution

As noted previously, scientists and philosophers have been vigorously debating the interpretations and implications of quantum theory since its inception. Heisenberg's matrix mechanics and Schrödinger's wave mechanics have had a revolutionary impact at the working level for scientists. Adoption of this formalism has resulted in what is arguably the most successful predictive scientific theory in human history. The replacement of theoretical working principles and their consequences for problems and practices of scientists has been compelling, justifying the revolutionary nature of quantum theory. One of the primary theoretical working principles of the pre-quantum theory era was the idea of the continuous nature of radiation, i.e. the wave picture of radiation.

Planck and Einstein changed the wave picture of radiation by demonstrating its corpuscular nature—Planck with his work on black-body radiation and Einstein with his work on the photoelectric effect. This then led to the suggestion of another opposing principle to the pre-quantum era—namely the wave nature of matter, which was proposed by de Broglie in 1923. Soon after, the formalism of Heisenberg and Schrödinger followed, resulting in new mathematical tools which scientists used to

[6]The perspective that we cannot know the world in itself and that we cannot know its essence per se goes back in Western thought at least to Kant (1787).

attack new problems—as well as old problems in new ways. Much of the important revolutionary scientific nature of quantum mechanics was provided by the probabilistic interpretation of Max Born. Experimental results were no longer seen as determined singular quantities, but as statistical probabilities of ensembles. This resulted in a substantial shift in how scientists approached and solved problems.

We also saw from previous treatments in Chap. 4 that the relationship of the observer to a system seems to be important. An important suggestion of the Copenhagen interpretation is that the observer is part of the total system to be considered for any given experiment. This brand of "holism" suggests that the observer cannot be separated from that which is observed. The uncertainty principle contributes in an important way to this view because it claims that due to the "interference" of the observer in the measurement process, there is always an uncertainty in our measurements or knowledge that cannot be overcome. As touched on prior, some versions of the Copenhagen interpretation suggest that the observer, by the act of measurement and/or observation, actually creates the reality observed.

The role of the observer is taken further with the many-worlds interpretation which suggests that the act of measurement or observation splits reality into different parallel worlds, all of which are actualized, but only one of which is consciously realized by any observer in that world. This is not part of the Copenhagen interpretations per se, but is similar to those views in the sense that the observer at least has some kind of correlation with a specific reality, if not a particular role in creating that reality. While these views solve problems, they also suffer from other problems that have motivated the consideration of alternatives, some of which were covered previously. These kinds of views can be accused of an extreme form of anthropocentrism—something that science has shown can be a misguiding tendency.

A claim of hidden variable interpretations is that there are unknown forces at work which do indeed completely objectively determine reality. The hidden variable program sees quantum theory as a kind of statistical mechanics. The motivation and claims of hidden variable theories seem reasonable even though such pursuits have sometimes been attributed to a desire for those who were too conservative to accept radical change (Squires 1994). The need for an observer as a vehicle for collapsing a wave function is eliminated since no collapse from an indeterminate state is needed to explain the results of a measurement. The critical question for this treatment is whether or not the uncertainty principle and its important implication for the role of the observer are consistent with hidden variable theories. The answer appears to be yes, and it is best captured by the following footnote in *Quantum Theory and Pictures of* Reality by Schommers: "For a realist it would of course be an elementary mistake to think the existence of hidden parameters is disproved by the uncertainty relations: for a priori the latter could conceivably just refer to limitations of possible human knowledge" (Schommers 1989, 102).

In other words, a hidden variable theory could coexist with the uncertainty principle which would indicate a limit to experimental measurements, while yet preserving a real, independent, deterministic world. For many, a deterministic, but epistemologically limited world is more palatable than "anti-realist" views or "observer-created" realities. Regardless, even the most conservative interpretations can be compatible with the uncertainty principle. This is important because the uncertainty principle suggests an important implication for the observer in obtaining knowledge, and to date, Heisenberg's principle still seems to hold. For Heisenberg, the uncertainty relations were the conceptual foundation on which all of quantum theory was built (Redhead 1990).

Even within the hidden variable program, the important role of the observer cannot be avoided. If the uncertainty principle holds, the observer disturbs the measurement, thereby limiting the accuracy of the measurement. Most fundamentally, the observer's perception of reality is limited by the presence of an observer as the vehicle for obtaining knowledge. The picture that emerges is that our understanding of reality is fundamentally limited because in some sense, observers "get in the way" from an otherwise direct access to ultimate reality. As Heisenberg was aware, this is a strong epistemological claim—and as was the case with relativity, this significant epistemological shift can be considered to be an important philosophical revolution.

As noted previously, the many-worlds interpretation was specifically referred to by Everett as a "relative state formulation." The relative state formulation suggests that completely independent states of observer and observed are not meaningful. Both are at least correlated or possibly "entangled," so the state of each must be specified and understood relative to each other. The correlation results from observations and measurements between subject and object and further shows how many treatments of quantum theory fundamentally *relate* the observer to that which is observed.

Relational quantum mechanics (RQM), covered previously, is similar to the many-worlds formulation in that it emphasizes that the state of a system describes how the observed system and the observer are related. But critically, as covered previously, RQM appears to be much more general because *all* physical interactions are part of this relative specification irrespective of the nature of the objects involved. A conscious observer is not needed. Physical interactions in general are the functional equivalent of a measurement in other interpretations. So, RQM is not just about objects and subjects per se, but about how *all* objects are *related* to each other. It is notable that many-worlds interpretations are similar to relational quantum mechanics in that they both claim a relational nature of all value assignments and properties of system. A key difference however is that many-worlds imagines a universal wave function that completely describes the universe as a whole, while RQM instead claims there is only a kind of web of interrelated incomplete "local" descriptions of the world. In this sense, RQM arguably implies epistemological limitations within our universe, while many-worlds implies fundamental epistemological limitations in measuring anything in other parallel worlds, or even detecting them at all.

It is worth taking a brief moment to highlight the potentially revolutionary nature of quantum entanglement—a very unusual area of quantum theory that has increasingly received empirical support (and will be touched on further in subsequent chapters). In a nutshell, quantum entanglement appears to suggest that particles affect each other at great distances—or at least that their quantum states are correlated or entangled in some way that allows some affect or "communication" between them, suggesting a form of interaction or relationship (Esfeld 2016) that apparently exceeds the speed of light—a kind of "non-locality" or "spooky action at a distance" as Einstein once put it. Spooky indeed. There is experimental recent evidence to suggest it does really happen (Freedman and Clauer 1972; Aspect et al. 1982; Hensen 2015) and there is tantalizing progress toward using quantum entanglement for communications (Yin 2017).

The word "entanglement" appears to have been coined by Schrödinger, and in 1935 he wrote a paper in which he captured the truly revolutionary nature of quantum entanglement and how fundamental it is to quantum mechanics: "I would not call entanglement *one*, but rather *the* characteristic trait of quantum mechanics, the one that enforces its entire departure from classical lines of thought." Entanglement was briefly touched on prior and will be touched on further to help support the general idea of a relational framework and the connection-action principle in Chap. 8.

5.3.5 Overturning Descartes

In the broadest sense, the shift from ontological theories to epistemological theories may be the most defensible shift brought by relativity and quantum theory. However, the treatment here has tried to show that the resulting substantive conceptual shift involved a transition to a form of observer-relatedness and potentially a broader form of relationality. The shift to epistemologically based physical theories implies some unusual and challenging consequences for observers in the pursuit of knowledge. So what specifically did this observer-relation revolution transition from? And why is it important?

Rene Descartes articulated a form of dualism which arguably dominated Western thinking until the twentieth century. Descartes' dualism claimed that the universe was made up of two fundamentally different things: mind and matter. Each had its own separate ontological status and occupied different ontological realms. This was a metaphysical framework by which Descartes and many others understood the world. The philosophical shift explored here was partly a shift from Descartes' dualism, which fundamentally separated the observer (the mind) from the observed (matter), to something closer to monism. Monism is suggested by relativity and quantum theory by incorporating the necessary role of the observer to have a fuller understanding of the single integrated system within which the observer resides.

This is an important change not only because a fundamental tenet has changed and because a methodological shift has occurred, but also because it recognizes and opens up a whole new area for philosophical and scientific study. This shift has brought the mind, the observer, into the pursuit of knowledge and reality in a way that appears to be critically important. We might consider the possibility that this shift has given birth to other revolutions such as the cognitive revolution (Baars 1986; Gardner 1987), which is what we would expect from a philosophical revolution as noted prior—i.e., the spin-off and applicability to other areas. It may also be that this shift has prompted the recognition of the probabilistic revolution (Krüger et al. 1987).

The observer-relation revolution appears to at least be an epistemological revolution—if not more. As touched on here, and to be explored further in later chapters, it may also hint at a broader revolution. It should be emphasized that despite language of revolution, this does not mean that the stronger version of "observer-dependence" (as opposed to something a little weaker such as "observer-relation") must have profound metaphysical or ontological implications. Physical theories can simply increasingly be more effective (e.g., increase their verisimilitude) by including observers as part of the system to be understood, possibly including the whole of universe, as the field of quantum cosmology is attempting to explore.

5.3.6 *Quantum Field Theory*

Another way to help further support the idea of the revolutionary and highly relational natures of relativity and quantum mechanics is to touch on quantum field theory (QFT). QFT is thought to be a kind of integration of quantum mechanics and special relativity (Weinberg 2005). QFT posits a world of fields in which disturbances in those fields (e.g., "waves," excitations, or "field quanta") can be thought of as the functional equivalent of particles. QFT helps provide an explicit description of photonic behavior, a "particle" that falls within the realm of special relativity since it is massless and travels at the speed of light. QFT might also ultimately explain broader fundamental force phenomena such as gravity in which a "graviton" is a "particle" (i.e., a field disturbance, a "wave," a "field quanta") that essentially causes and mediates the force of gravity.

Fields have been an important construct in many domains of science, including much of "classical physics" such as electromagnetic fields and Newtonian gravitation. However, as is often the case, trying to intuitively comprehend a "field" in terms of what it *really is*, particularly in material terms that our minds are accustomed to, is illusive. Nevertheless, fields can be seen to be quite physically real in many ways. At the very least, the *effects* of many fields are quite evident, for example, with a magnetic field or a gravitational field. In the sense that we can directly see the effects of fields, they are presumably physically real and critically active in the world. It may be that the mathematical constructs fall short of capturing the "essence" of what a field is, or the essence of ultimate reality, but the *action* of fields in the world certainly seems real.

Regardless of whether fields are real in the way we would like them to be, it is notable for our purposes here that while fields are "quantized" in quantum field theory to help capture the behavior of "particles" as localized field quanta, the field picture nevertheless suggests that there is not really a particle in the discrete material way it is often conceived. Instead, particles can be thought of as manifestations of field excitations, as disturbances within a field, including when different fields interact. So, fields may not be "real" in a more satisfying way that is traditionally or intuitively materialist or substantive. What we can say is that fields at least appear to *specify relations*. Representations of fields are at least relational in the sense that they specify relational dynamics within a field (e.g., excitations that represent particle behaviors) and external to a field (e.g., interactions between fields). In this sense, fields may be better thought of as describing a "relational reality," and fields themselves may ultimately simply be relational "entities" without substance in the usual materialist way of conceiving reality.

A relationalist picture of fields can be further developed by suggesting that quantum fields, while locally quantized to capture certain behaviors, are still ultimately fully and completely interconnected, or possibly *intra*-connected, and effectively "infinite" fields permeating all of reality. Fields can be thought of as tightly integrated constructs that may indeed be indicative of the very tightly integrated, highly connected, and relational nature of reality. Compared to more traditional, discrete, atomic, materialist conceptions of reality, a field then is arguably a more continuous and more "seamlessly" integrated picture of reality. A field-based picture of physical reality can be interpreted to suggest a high degree of seamless connectedness and relatedness, or "relationality," in the physical world.

5.3.7 Different Kinds of Relationalism?

One might object that the kinds of relationalism in relativity and quantum theory are different enough so as not to warrant grouping them together into a broader philosophical revolution.[7] Perhaps this is partly why Einstein took the stand he did against quantum theory being a complete theory. Entanglement and the limitations of knowledge and other associated implications arrived at by considering the role of the observer were too extreme compared to the observer-dependence in his theory of relativity. In some ways, Einstein's relativity is also a theory of absolutes—e.g., the speed of light and the laws of physics. Both are invariant. This gave Einstein some absolutes. Such absolutes can imply an independent reality. Many of the implications of quantum theory however seemed to imply otherwise—especially Bohr's Copenhagen interpretation. Einstein's reaction may be seen as a way for him to counter indeterminism, resulting in an underestimation of the importance of the relation to an observer that even his own theory reveals. We might not be too surprised by this since a truly revolutionary shift can run deep enough so as to elude even those who help bring it about.

Also, if any version of a Copenhagen interpretation is correct, the observer-relatedness of relativity and quantum theory can be seen as a difference of degree as opposed to a difference of kind. Relativity clearly shows that the state of an observer contributes substantively to what is measured and hence "known." The Copenhagen

[7] Bub (2016, pp. 6, 7) notes that relativity and quantum theory are "analogous" revolutions in that they both represent "a fundamental structural change in the way we represent how events fit together." He makes the distinction that relativity is a spatiotemporal restructuring and quantum theory is an informational restructuring. It is notable for the purposes of this chapter that at least special relativity can also be seen as an informational restructuring as well, given that it rests importantly on assumptions about the limits of transmitting information via the speed of light.

interpretation of quantum theory suggests that the observer plays an important role in creating what is known. Both are epistemological claims. But some versions of the Copenhagen interpretation arguably begin to approach a metaphysical claim. Regardless, in both theories, the relationship of the observer to observed plays an important role in what is ultimately known. In many interpretations of quantum theory, the role of the observer is greater.

However, even if we were to consider that there is a difference in "kind" of observer-dependence (say, if we include the realist "hidden variable/knowledge limit" interpretation) we can still argue the case for an observer-relation revolution via an argument suggested above—namely, that the theories are data points for a broader philosophical set of implications. In this case, we have two kinds of observer-relation or relational epistemology more generally. For relativity, we might call it "value measurement relationalism." For quantum theory, we might call it "knowledge limit relationalism," both of which can be considered forms of epistemological relationalism. So, we can still infer observer relationalism by considering that both kinds of relationalism contribute to the overall general philosophical shift.

5.3.8 A Relational Synthesis for Relativity and Quantum Theory

Quantum theory appears to be a theory that critically includes us, observers. Knowledge of reality, and arguably reality itself, cannot be fundamentally separated from the observer.[8] If stronger versions of the Copenhagen interpretation are correct, reality in some sense depends on the observer. If the hidden variable interpretation is correct, then predictive knowledge may be fundamentally limited because of observer-dependent limitations. If many-worlds is correct, observers and "measurements" create alternative parallel universes, alternative realities. If relational quantum mechanics is correct, we take a large step beyond the relationships of observers to the world, to a much more general suggestion that all interactions and measurements are relative to all other objects regardless of what those objects are. Minimally, quantum mechanics appears to suggest a very strong form of relational epistemology.[9]

Both relativity and quantum mechanics suggest an important relationship between the observer and observed, and both essentially eschew the idea of absolute frames of reference. Relational quantum mechanics goes further, removes any privileged role for an observer, and simply suggests that what is knowable and represented by quantum states is only relations and interactions in a system. Any object or system is

---

[8]Another interpretation called quantum Bayesianism, or "QBism," goes so far as to suggest that wave function probabilities should be understood only as a single individual observer'ssubjective state of knowledge of a system at any given moment (Fuchs et al. 2014) and so is different from Copenhagen interpretations in part because the wave function applies to the experience of only a particular observer, agent, or "user," which is different from that of another user (Mermin 2017).

[9]Healey (2012) covers a variety of ways in which quantum states can be thought of as "relational" and emphasizes that any such relationality does not necessarily imply subjectivity.

knowable only in the way it is related to some other object or system which can act as a frame of reference, including, but not limited to, conscious observers. This appears to at least be a form of "epistemic structural realism" where our knowledge is limited to the structures of relationships and does not make ontological claims (Ladyman 1998). "Ontic structural realism" however does make stronger ontological claims about the realism of relations and relational structures more generally (French and Ladyman 2011, Morganti 2011), and RQM has been interpreted in this light (Dorato 2016, Candiotto 2017).

Rovelli (2004) extrapolates RQM further and explores potential connections between quantum theory relationalism and the spatiotemporal relationalism of general relativity, again suggesting something fundamental about "relationality" more generally. And Vidotto (2017) goes further to suggest that the relational understanding of quantum mechanics has the possibility to lead to a coherent relational quantum cosmology.

Quantum field theory (QFT), which leverages the field construct of general relativity and the quantized construct of quantum mechanics, can also be viewed as suggesting deep degrees of connectedness and relationality in the physical world—as touched on prior. The notion of a field, or quantum field, is a singular "intra-connected" construct, or structure, that fundamentally drives and manifests physical dynamics and has also been interpreted as a form of ontic structural realism (Lyre 2004). And notably, QFT can be traced to a third major early formulation of quantum theory (the first two coming from Schrödinger and Heisenberg) which was first developed by Paul Dirac and has been called the "*interaction* picture" (Sakurai and Napolitano 2017)—a name that has further relevance in the subsequent chapter on the connection-action principle.

While there are some connecting threads between relativity and quantum theory (Nugaev 1988), they are essentially theories of different domains, which partially accounts for their independent development. But because these theories arose independently to some extent, and yet share the common heritage of causing a similar ontological-to-epistemological shift, we might gain confidence about that shift from the independent convergence. If there is reason to think the theories can in principle be completely merged as quantum field theory has indicated,[10] suggesting some deeper level of compatibility, we can have additional confidence in the independent convergence that would even more strongly support relational implications. Both theories did not start out attempting to reveal observer-relation implications, but each came to independently reveal the importance of the relationship between observers and the rest of world; i.e., they both revealed the deep relational nature of knowledge and, in the cases of certain interpretations such as RQM and QFT, gesture at a deeply relational world more generally.

[10]The successes of quantum electrodynamics and quantum chromodynamics are examples and may be indicative of a trend that could be continued by finding a quantum field theory of gravity.

5.3.9 Science and Philosophy

For the purposes of the scientific–philosophical distinction explored here, it may be argued that in the case of quantum theory, observer-relatedness does have direct operational or working significance for scientists, perhaps then reducing its philosophical implications. Heisenberg himself could be cited as an example. The "measurement problem" has been examined by many, but it is difficult to assess the extent to which it has had operational or working significance for practicing scientists.

Karl Popper states that "most physicists who quite honestly believe in it [the Copenhagen interpretation] do not pay any attention to it in actual practice" (Popper 1967, 8). Even if one were to successfully argue that the measurement problem has direct working applicability for the scientist, we must remember that observer-relatedness is a more general principle derived partly from the ideas revealed by the measurement problem. This may be an example of the necessary and expected blurring between epistemology and science. Quantum mechanics may be using, or transforming, observer-relatedness into a scientific concept. The idea that observer-relatedness implies that knowledge depends in part on the observer is general enough to potentially have extremely wide applicability, some of which we will explore further.

A trend in knowledge has often been that those questions of a philosophical or metaphysical nature eventually become scientific questions. In this way, the distinction between philosophy (e.g., epistemology and metaphysics) and science may be helpful to point the way to new frontiers for science to examine. This does not mean however that there is a domain for the philosopher and a separate domain strictly for the scientist. If we take Lakatos' naive falsificationist definition that "metaphysical" assertions have no potential falsifiers, such a claim of separate domains existing for the scientist and philosopher might be justified. However, Lakatos' definition has a severe difficulty. How are we to know in all cases with a high level of confidence that a claim has no potential falsifiers? Even his use of the word "potential" gives it away. It is very difficult, if not impossible, to know with certainty that a claim absolutely cannot be falsified either at the present, or more importantly, in the future. Regardless, too strongly demarcating between domains of science and philosophical considerations could be misguided and perhaps even counterproductive. The efforts of philosophy and science can help inform each other, and in some cases, each merges with the other (McMullin 1981).[11]

5.4 From Biology to Culture

Here, we will very briefly explore the basics of biological, psychological, and cultural evolution to illustrate their relational nature, focusing primarily (though not

[11] Ernan McMullin emphasizes that a "philosophic issue is a philosophic issue whether it be discussed by a philosopher or a scientist" (McMullin 1981, 181).

exclusively) on what is arguably a weak sense of relationalism in the form of causal relations (see Wildman 2010 for an exploration of this and other kinds of distinctions for relationalism). This can help point to stronger or deeper notions of relationalism that will be explored further in subsequent discussions and in the development of more specific worldviews that follow.

5.4.1 *Biological Evolution*

Before human beings learned about cosmic evolution, we learned about our biological evolution. Charles Darwin helped show us how we developed as a species, how all life on Earth evolved. We learned that a simple process of natural selection—or, as noted in Chap. 3, what might be thought of even more simply as "filtered replication"—has given rise to the extreme diversity of life on Earth today. Part of what is noteworthy about biological evolution via natural selection is how it is so tightly causally related to "internal" states, the environment, and the universe at large. Biological systems are built from molecules born in the furnaces of stars. The young dynamic Earth, along with cosmic sources of energy and radiation, mixed and mashed molecules to form molecular replicators that slowly evolved over time to form the remarkable array of life we see today.

While it may seem trivial to emphasize the critical relationship of organisms to their environment, it does seem worth noting for our purposes that biological systems must in some sense tightly relate to, or tightly "fit" into their broader environment in order to survive and replicate. Natural selection does indeed imply that nature, or the environment, in some broad sense, "selects" organisms that are best able to exist and reproduce in the particular environment, including other living things in the environment. But as noted in Chap. 3, the word "selection" is somewhat misleading. It is not selection in the intentional sense of course. It is instead a kind of filtering based primarily on how an organism is related to everything around it, as well as its internal relations. If the organism is well suited to survive and replicate in the context of its web of external and internal relations, it will persist and replicate over time. And in general, an organism must be better at surviving and replicating than other organisms in the environment that it competes with. The broader web of relations in which organisms find themselves is critical to their physical structure, behavior, survival, and reproduction, and there are increasing attempts to model these complex relationships with an overall relational emphasis and relational modeling approaches (Kineman 2011).

5.4.2 *Cognitive Evolution*

An important adaptation resulting from biological evolution is a form of general-purpose intelligence that allows for highly dynamic responses (e.g., rapid and novel problem solving and learning) to highly complex and dynamic environments, including social groups (Dunbar 1998). The evolution of intelligence of many species has been a key element for survival and reproduction, with *Homo sapiens* being a notable example. Human intelligence has arguably been driven by social selection pressures that require unusually complex "computations" and problem-solving abilities, including those abilities that allow tribes or groups to behave effectively as groups.

Group cohesion, arguably a key result of biological evolution and sociality, along with somewhat flexible brains of modern humans, has given rise to cultural evolution to allow groups to learn and act as effective groups. Group success, like evolution and selection at other levels, is critically driven by environmental and other selective pressures, including how group members relate internally and externally with other groups. The relationships of the individual to its broader environment, perhaps most importantly the social environment, have presumably shaped the architectures of human brains. Relational models for thinking about human interaction and human psychology continue to be explored and appear to have promise (Gergen 2009).

5.4.3 Cultural Evolution

Culture can generally be thought of as a set of strategies and constructs for understanding our world and effectively living in it, including our social world. For this and subsequent discussions, it will be helpful to think more specifically about culture as the *collective manifestation of value*—where value is that which is valuable to "sufficiently complex" agents, from which meaning, purpose, ethics, and aesthetics can be derived (Lupisella 2009). This is very similar to the pursuits of normative aspiration touched on previously. Culture manifests value in many varied forms, from thoughts and knowledge, to symbolic abstractions and social norms, to mass movements and large-scale physical creations. Cultural evolution then is the change in culture over time.

There is much wiggle room in this characterization of culture to accommodate a variety of perspectives about culture. For example, memetic perspectives of evolution may see mere replication as a manifested "value," but memes ("replicating cultural units" such as ideas, art, ways of doing things, etc.) would not necessarily be thought of as having autonomy in the sense of being able to consciously "choose" a behavior to ensure their replication. Nor do memes have to be seen necessarily as providing usefulness to agents (Blackmore 2009). Reference to "sufficiently complex beings" also offers wiggle room in that it could, based on some admittedly loose interpretations, include things like bacterial colony behavior (Bloom 2009). But if collective bacterial behavior is not thought to be sufficiently complex and/or sufficiently intentional, it would not count as culture—whereas collective human behavior appears to be far more complex and seemingly highly intentional and so would be considered culture. Cultural evolution then can be seen as the change in culture over short and long timescales, requiring integrated interdisciplinary approaches (Richerson and Boyd 2005; Denning 2009; Kolodny et al. 2018).

Modern cultural evolution is a unique and compelling phenomenon, in part because it arguably has the potential to leave biological evolution behind—and is probably doing so already. For sure, biological impulses exert tremendous and often blinding influence on human behavior and human culture more broadly, but the level of awareness exhibited by human beings in the modern world is still noteworthy. Another notable feature of cultural evolution that partly contributes to our awareness is the sheer number of seemingly ever-increasing complex relationships evolving over time—well beyond interpersonal relationships (e.g., including the evolution of technology). Human cultural evolution is exhibiting a kind of relational complexity and power unknown in our planet's history. Human culture now includes such a wide variety of activities with such rapid rates of growth that we have to take seriously the possibility that cultural evolution may someday have, if it does not already, a kind of cosmic relevance—a notion that will be explored subsequently in the chapter on cosmocultural evolution.

5.4.4 Cosmic Evolution: Complexity and Self-organization?

> The evolution of the Universe is a learning process.
>
> —Corliss (1989)

From our explorations of scientific minimalism and the relationalist framework being explored here, we see not only that the universe has been evolving, but that its evolution seems to include important relational qualities. Despite assertions that evolution is not directional or, "progressive", it does not seem unreasonable to contemplate possibilities for trends given some of the long-term dynamics of cosmic, biological, and cultural evolution. For example, there has been an increasing recognition of the possibility that the universe has been steadily evolving toward increasing degrees of "complexity" (Davies 1995; Chaisson 2005, 2009; Christian 2009). One simple definition of complexity theory notable for the purposes of this book is "the study of the phenomena which emerge from a collection of interacting objects" (Johnson 2007, 3).

Regarding complexity in the universe as a whole, Paul Davies writes:

> I believe that there's a sort of 'law of increasing organized complexity' operating in the universe. It's not quite a law in the same sense as, say, Newton's law of gravity, more like a *tendency* or *trend* (italics added), but its manifestation seems unmistakable. There really does seem to be a general tendency in nature for increasing organizational complexity (or depth). (Davies 1995, 105)

Carroll (2016) acknowledges the possibility of a law, or at least a trend toward increasing complexity, but also cautions that we do not really have a precise understanding of complexity (indeed, he talks of the "apparent complexity" that he and his colleagues have studied). Perhaps more perhaps importantly, Carroll notes that while there is reason to think that complexity can emerge and grow over time, it need not do so *necessarily*. Equally importantly, any complexity that does emerge in the universe is still very likely to be temporary, eventually returning to a simpler state—due largely, if not completely, to the second law of thermodynamics. Carroll notes that cream sitting "simply" on the top of coffee can transition to a state of complexity as entropy increases and the cream mixes with the coffee, eventually returning to a simple state of uniform mixing (a "full blend" of sorts). He writes:

> Those swirls in the cream mixing into the coffee? That's us. Ephemeral patterns of complexity, riding a wave of increasing entropy from simple beginnings to a simple end. We should enjoy the ride. (Carroll 2016, 236)

Notably, for the purposes of this book (which will become more apparent later), Carroll points out that while this "coffee cup complexity" does not have to necessarily emerge—rather, its emergence is dependent on how the cream and coffee *interact* with each other, or more generally, how they relate to each other.[12]

If a potential trend or law of increasing complexity and/or self-organization for the universe as a whole is going too far, then we might at least consider "increasing complexity" as a trend for just biological and cultural evolution, along with its philosophical and ethical relevance—which will be touched on further in subsequent sections. Paul Davies notes that life and consciousness can be seen as *emergent* properties, as "…something that emerges in a physical system when it reaches a certain level of complexity" (Davies 1995, 98). A similar idea is suggested by Stuart Kauffman to explain the origin of life emerging as a phase transition of a relatively complex network of chemical interactions (Kauffman 1995).[13]

Similar to the complexity noted by Carroll and Kauffman, an emergent property is more about the relationships and interactions, the functional organization, or perhaps the information content, of any given system—as opposed to merely what its substantive content is or what it is "made of." Biological evolution does seem to have created increasingly complex organisms and collections of organisms. And biological evolution has led to cultural evolution, which appears to be a staggeringly complex and rapid form of evolution with tremendous power—including the power to control and modify the physical environment precisely and on very large scales. Such complexity embodies extreme manifestations of relationships (in form and amount) across our globe and increasingly into our wider universe as humanity, and potentially other beings, expand into space.

5.5 Summary of Chapter 5: A Relational Framework

In this chapter, we explored some philosophy and science to help lay a foundation for a relational way of thinking, a relationalist framework, or possibly even a relational

---

[12] I now enjoy cups of coffee even more!

[13] Heylighen et al. (2015) propose "chemical organization theory" as a universal modeling framework, leveraging an "action ontology" as a starting point—a topic we will touch on further in Chap. 8.

worldview. We noted the relational nature of process philosophy and explored the relationalist "revolutions" of relativity and quantum theory. We briefly noted relational aspects of biology, psychology, and cultural evolution (including complexity) and suggested that the complexity and potential self-organization of cosmic evolution can be seen as a highly complex web of relations. As Fig. 1.1 shows, this relationalist context can serve as a broad framework within which to think about stronger versions of scientific minimalism (e.g., that acknowledge a key role for physical relationships and physical dynamics) and to help inform cosmological theories of value and more philosophical worldviews such as cosmological reverence, cosmocultural evolution, and the connection-action principle—all of which will now be developed in Part II.

References

Aspect, A., Grangier, P., & Roger, G. (1982). Experimental realization of Einstein-Podolsky-Rosen-Bohm Gedanken experiment: A new violation of Bell's inequalities. *Physical Review Letters, 49*(2), 91–94.

Baars, B. J. (1986). *The cognitive revolution in psychology*. New York: The Guilford Press.

Blackmore. (2009). Dangerous memes; or, what the Pandorans let loose. In S. J. Dick & M. L. Lupisella (Eds.), *Cosmos and culture: Cultural evolution in a cosmic context*.

Bloom, H. (2009). The big burp and the multi-planetary mandate. In S. J. Dick & M. Lupisella (Eds.), *Cosmos and culture: Cultural evolution in a cosmic context*. Washington, DC: NASA History Series.

Bub, J. (2016). *Bananaworld: Quantum mechanics for primates*. Oxford University Press.

Burkhart, B. Y. (2004). What Coyote and Thales can teach us: An outline of American Indian epistemology. In A. Waters (Ed.), *American Indian thought*. Mauldin: Blackwell Publishing.

Candiotto, L. (2017). "The Reality of Relations", *Giornale di Metafisica, 2017*(2): 537–551.

Carroll, S. (2016). *The big picture: On the origins of life, meaning, and universe itself*. New York: Dutton.

Chaisson, E. (2005). *Epic of evolution: Seven ages of the cosmos*. New York: Columbia University Press.

Chaisson, E. (2009). Cosmic evolution: State of the science. In S. J. Dick & M. Lupisella (Eds.), *Cosmos and culture: Cultural evolution in a cosmic context*. Washington, DC: NASA History Series.

Chen, S., Boucher, H., Kraus, M. (2011). The relational self. In S. J. Schwartz, K. Luyckx, V. L. Vignoles (Eds.), *Handbook of identity theory and research* (pp. 149–175). Springer.

Christian, D. (2009). History and science after the chronometric revolution. In S. J. Dick & M. Lupisella (Eds.), *Cosmos and culture: Cultural evolution in a cosmic context*. Washington, DC: NASA History Series.

Cohen, I. B. (1985). *Revolution in science*. Cambridge: Harvard University Press.

Corliss, J. B. (1989). The flow of energy, natural learning systems and the creation of life on Earth. *Acta Astronautica, 19*(11), 869–873.

Davies, Paul. (1995). *Are we alone?*. New York: HarperCollins.

Denning, K. (2009). Social evolution. In S. J. Dick & M. Lupisella (Eds.), *Cosmos and culture: Cultural evolution in a cosmic context*. Washington, DC: NASA History Series.

Dijksterhuis, E. J. (1950). *The mechanization of the world picture, 1961* (2nd ed.). Oxford: Oxford University Press.

Dorato, M. (2016). Rovelli's relational quantum mechanics, Anti-Monism, and quantum becoming. In A. Marmodoro & D. Yates (Eds.), *The Metaphysics of Relations* (pp. 235–262). Oxford: Oxford University Press.

Dunbar, R. (1998). The social brain hypothesis. *Evolutionary Anthropology: Issues, News, and Reviews, 6*(5), 562–572.

Esfeld., M. (2016). The reality of relations: The case from quantum physics. In A. Marmodoro & D. Yates (Eds.), *The Metaphysics of Relations* (pp. 218–234). Oxford University Press.

Freedman, S. J., & Clauser, J. F. (1972). Experimental test of local hidden-variable theories. *Physical Review Letters, 28*(14), 938–941.

French, S. & Ladyman, J. (2011). In defence of ontic structural realism. In A. Bokulich & P. Bokulich (Eds.), *Scientific Structuralism* (pp. 25–42).Dordrecht: Springer Netherlands.

Fuchs, C., Mermin, D., & Schack, R. (2014). An introduction to QBism with an application to the locality of quantum mechanics. *American Journal of Physics, 82*, 749. https://doi.org/10.1119/1.4874855

Gardner, (1987). *The mind's new science: A history of the cognitive revolution*. New York: Basic Science.

Gergen, K. J. (2009). *Relational being: Beyond self and community*. New York: Oxford University Press.

Goldberg, S. (1984). *Understanding relativity: Origin and impact of a scientific revolution*. Birkhäuser Boston.

Hartshorne, C. (1971). The development of process philosophy. In E. H. Cousins (Ed.), *Process theology*. New York: Newman Press.

Healey, R. (2012). Quantum theory: A pragmatist approach. *The British Journal for the Philosophy of Science, 63*(4), 729–771. https://doi.org/10.1093/bjps/axr054.

Hensen, B., et al. (2015). Loophole-free Bell inequality violation using electron spins separated by 1.3 kilometres. *Nature, 526*, 682–686.

Heylighen, F., Beigi, S., & Veloz, T. Chemical organization theory as a universal modeling framework for self-organization, autopoiesis and resilience (ECCO Working Papers No. 2015-01). Retrieved from http://pespmc1.vub.ac.be/Papers/COT-applicationsurvey.pdf.

Ingram, P. (2006). *Constructing a relational cosmology*. Princeton Theological Monograph Series 62. Pickwick Publications.

Johnson, N. (2007). *Simply complexity: A clear guide to complexity theory*. One World Publications.

Kaipayil, J. (2009). *Relationalism: A theory of being*. Bangalore: JIP Publications.

Kant, I. (1787). *Critique of pure reason* (Trans. F. Muller, 1981), (2nd ed., 1896). The Macmillan Company.

Kauffman, S. (1995). *At home in the universe*. Oxford: Oxford University Press.

Kineman, J. J. (2011). Relational science: A synthesis. *Axiomathes, 21*(3), 393–437 (Springer).

Kitcher, P. (1993). *The advancement of science*. Oxford: Oxford University Press.

Kolodny, O., Feldman, M., & Creanza, N. (2018). Integrative studies of cultural evolution: Crossing disciplinary boundaries to produce new insights. *Philosophical Transactions of the Royal Society B, Biological Sciences, 373*(1743).

Kraus, E. (1998). *The metaphysics of experience: A companion to whitehead's process and reality*. Fordham University Press.

Krüger, L., Daston, L. J., & Heidelberger, M. (Eds.). (1987). *The probabilistic revolution*. Cambridge: The MIT Press.

Kuhn, T. S. (1957). *The Copernican revolution*. Cambridge: Harvard University Press.

Kuhn, T. S. (1962). *The structure of scientific revolutions* (2nd ed., 1970). Chicago: University of Chicago Press.

Ladyman, J. (1998). What is structural realism? *Studies in History and Philosophy of Science Part A, 29*(3), 409–424.

Lakatos, I. (1978). *The methodology of scientific research programmes* (J. Worall & G. Currie (Eds.)). Cambridge University Press.

Laudan, L. (1977). *Progress and its problems*. Berkeley: University of California Press.

Lucas, G. (1989). *The rehabilitation of whitehead: An analytic and historical assessment of process philosophy*. Albany: State University of New York.

Lupisella, M. (2009). Cosmocultural evolution: The coevolution of cosmos and culture and the creation of cosmic value. In S. J. Dick & M. L. Lupisella (Eds.), *Cosmos and culture: Cultural evolution in a cosmic context* (pp. 321–359). NASA SP-2009-4802. http://history.nasa.gov/SP-4802.pdf.

Lupisella, M. (2019). Is the universe enough? Can it suffice as a basis for worldviews? *Journal of Big History, III*(3), 123–140.

Lyre, H. (2004). Holism and structuralism in U(1) gauge theory. *Studies in History and Philosophy of Science Part B: Studies in History and Philosophy of Modern Physics, 35*(4), 643–670.

Marmodoro, A. and Yates, D. (Eds.). (2016). *The Metaphysics of Relations*. Oxford: Oxford University Press.

McAllister, J. (1996). *Beauty and revolution in science*. Cornell: Cornell University Press.

McMullin, E. (1981). Is philosophy relevant to cosmology? *American Philosophical Quarterly, 18*(3).

Mermin, N. D. (2017). Why QBism is not the copenhagen interpretation and what John Bell might have thought of it. In R. Bertlmann & A. Zeilinger (Eds.), *Quantum [Un]speakables II. The frontiers collection*. Springer International. https://doi.org/10.1007/978-3-319-38987-5_4.

Mesle, R. (2008). *Process-relational philosophy: An introduction to Alfred North Whitehead*. Templeton Foundation Press.

Meyer, S. (2005). "Introduction" in "Whitehead now" a special issue of *Configurations* 1(13), 1–33 (S. Meyer & E. Wilson (Eds.)).

Moore, A. D. (2004). Values, objectivity, and relationalism. *The Journal of Value Inquiry, 38*, 75–90.

Morganti, M. (2011). Is there a compelling argument for ontic structural realism? *Philosophy of Science, 78*(5), 1165–1176.

Newton-Smith, W. H. (1981). *The rationality of science*. London: Routledge and Kegan Paul.

Nugaev, R. N. (1988). Special relativity as a stage in the development of quantum theory. *Historia Scientarium No. 34*.

Popper, K. R. (1967). Quantum mechanics without 'the observer.' In M. Bunge (Ed.), *Studies in the foundations methodology and philosophy of science* (Vol. 2).

Redhead, M. (1990). Quantum theory. In R. C. Olby, G. N. Cantor, J. R. R. Christie, M. J. S. Hodge (Eds.), *Companion to the history of modern science*.

Rescher, N. (1996). *Process metaphysics: An introduction to process philosophy*. State University of New York Press.

Richerson, P. J., & Boyd, R. T. (2005). *Not by genes alone: How culture transformed human evolution*. Chicago: University of Chicago Press.

Rovelli, C. (2004). *Quantum Gravity (Cambridge Monographs on Mathematical Physics)*. Cambridge: Cambridge University Press.

Rukeyser, M. (1968). The speed of darkness from *The speed of darkness*, Random House and from *Out of silence: Selected poems* (K. Daniels (Ed.)) (1992). Evanston, Illinois: TriQuarterly Books, Northwestern University Press.

Sakurai, J. J., & Napolitano, J. (2017). *Modern quantum mechanics* (2nd ed.). Cambridge University Press.

Schlosshauer, M. (Ed.). (2011). *Elegance and enigma: The quantum interviews*. Berlin: Springer.

Schommers, W. (1989). *Quantum pictures and theories of reality*. Berlin: Springer.

Schrödinger, E. (1935). Discussion of probability relations between separated systems. *Mathematical Proceedings of the Cambridge Philosophical Society, 31*(4), 555–563.

Spretnak, C. (2011). *Relational reality: New discoveries of interrelatedness that are transforming the modern world*. Topsham: Green Horizon Books.

Squires, E. (1994). *The mysteries of the quantum world*. Philadelphia: Institute of Physics Publishing.

Stachel, J. (1990). The theory of relativity. In R. C. Olby, G. N. Cantor, J. R. R. Christie, & M. J. S. Hodge (Eds.), *Companion to the history of modern science*.

Thagard, P. (1992). *Conceptual revolutions*. Princeton: Princeton University Press.

Vidotto, F. (2017). Relational quantum cosmology. In K. Chamcham, J. Silk, J. Barrow, & S. Saunders (Eds.), *The philosophy of cosmology* (pp. 297–316). Cambridge: Cambridge University Press.

Weinberg, S. (2005). *Quantum theory of field*. Cambridge University Press.

Whitehead, A. N. (1929). *Process and reality: An essay in cosmology* (Edition 1978 by D. R. Griffin & D. W. Sherbourne), New York: Macmillan.

Wildman, W. (2010). An introduction to relational ontology. In J. Polkinghorne (Ed.), *The trinity and an entangled world: Relationality in physical science and theology*. USA: Wm. B. Erdmans Publishing Company.

Yin, J., et al. (2017). Satellite-based entanglement distribution over 1200 kilometers. *Science, 356*(6343), 1140–1144.

Young, M. (2011). Relevance and relationalism. *International Journal for Ontology and Metaphysics, 12*, 19–30 (Springer).

Part II
Cosmological Worldviews and Implications

Chapter 6
Cosmological Reverence

6.1 Background and Overview

As we move from left to right in Fig. 1.1 (shown in Chap. 1), we see that "stronger" versions of scientific minimalism, particularly in the context of a broader relationalist framework previously explored, can overlap with and lead to worldviews that might be characterized under an umbrella term of "cosmological reverence." *Cosmological reverence* can be thought of as a view in which our relationships to the universe as a product of and ongoing manifestation of cosmic evolution are thought to be important and valuable to the point of justifying some degree of reverence for the universe, and in particular, for its evolution. It is not simply a reverence for the universe in a general sense, but more specifically for the *evolutionary* dynamics revealed by modern cosmology. It is a reverence for the unfolding processes of cosmic creativity, and for the long view of cosmic evolution..

Scientific minimalism is shown in Fig. 1.1 as overlapping with the overall relationalist framework because scientific minimalism, as per details covered in Chap. 4, can easily acknowledge and emphasize the fundamental importance of relations and relationality more generally. Scientific minimalism also overlaps a bit with cosmological reverence, in part because it could be argued that stronger versions of scientific minimalism could still acknowledge a certain kind value associated with our relationships to the universe—something that cosmological reverence emphasizes more strongly.

Indeed, cosmological reverence suggests much stronger assertions of value and can be seen as a kind of "cosmological theory of value"—which can then inform

---

Parts of this chapter are taken from a previous book chapter from M. Lupisella, "Cosmocultural Evolution: The Coevolution of Cosmos and Culture and the Creation of Cosmic Value," in *Cosmos and Culture: Cultural Evolution in a Cosmic Context* (2009), eds. Steven J. Dick and Mark L. Lupisella. Additional material taken from M. Lupisella, "Life, Intelligence, and the Pursuit of Value in Cosmic Evolution," in *The Impact of Discovering Life Beyond Earth*, (2015) ed. Steven. J. Dick. Cambridge University Press, based on U.S. Library of Congress workshop in November 2014.

M. Lupisella, *Cosmological Theories of Value*, Space and Society,
https://doi.org/10.1007/978-3-030-25339-4_6

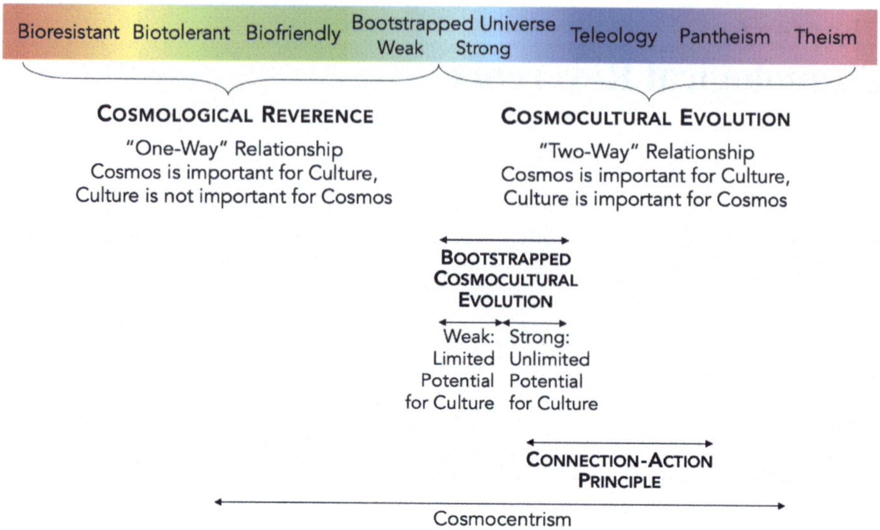

Fig. 6.1 Relationships between cosmos and culture

explorations of meaning, normative aspiration, cosmic ethics, and perhaps even "cosmocentric" ethical frameworks. We will apply cosmological reverence to a few issues such as the search for, discovery of, and potential interactions with, extraterrestrial life and intelligence.

Figure 1.1 places cosmological reverence in an overall relationalist framework. Figure 6.1 places cosmological reverence in another broad context that explores potential perceptions of relationships between cosmos and culture. Figure 6.1 shows that cosmological reverence captures what can be seen as "one-way" or "unidirectional" relationships between culture and cosmos in the sense that the universe is important for culture, but culture is not important for the universe.

In cosmological reverence, cultural beings evolve from and are sustained by the universe, and may revere the universe—but cultural beings and cultural evolution do not have significance for the universe as a whole. In such "unidirectional" relationships, culture is strongly influenced by the universe because culture is created by, informed by, and ultimately limited by physical laws and cosmological realities. As touched on previously, human culture is imbued with a wide variety of imaginative and influential worldviews, literature, music, and other forms of culture that are directly and profoundly influenced by cosmological perspectives (e.g., Swimme and Berry 1992; Dick 2000; Palmeri 2009; Vakoch 2009; Carroll 2016).

6.2 The Role of Biology

Even with a "bioresistant universe" (shown at the left in Fig. 6.1) that suggests life emerged against substantial odds, there could still be an important recognition of life's relationship to the universe as an outcome of cosmic evolution—however random and unlikely it may have been. A broader version of this kind of view might also hold that the origin of the universe itself was essentially random, and that randomness is an important factor in the origin and evolution of the universe. Nevertheless, even this kind of "accidental universe" worldview cannot escape the fact that cosmology influences culture in an important way by influencing worldviews and constituting important pillars of scientific culture and all that implies for our broader culture. It is also a view that can still recognize value in our "one-way" relationships with the universe more broadly in that we have originated from, and are in some sense sustained by, the broader universe.

A "biotolerant universe" suggests that life is highly contingent in the sense that laws of the universe allow that life, and perhaps eventually intelligence and culture, *may* originate under the proper, perhaps narrow, set of conditions, e.g., the presence of liquid water, a dynamic environment driving evolution, etc. This view sees the characteristics of the universe as allowing life to emerge and perhaps intelligence to follow, but the universe's laws and constants are not such that life is an expected outcome of cosmic evolution—or a "cosmic imperative" as some like to say. Life may arise periodically, and intelligence and culture may then arise periodically (and perhaps even frequently given the large number of potential environments for life and intelligence to evolve), but this can happen without any particular "inevitability" or significance for the universe as a whole.

A "biofriendly universe" suggests that the nature of the universe tends to produce life. Replication, self-organization, and life arise "easily" and perhaps often as a result of the universe's laws, physical constants, and prevalent conditions throughout the universe. An example of this kind of view might also be called a "biophysical cosmology" or "biological universe"—a popular view based on both philosophical and scientific grounds that has been explored by a wide variety of philosophers and scientists (Dick 1996; Dick and Strick 2004). However, regardless of how predisposed the universe may be to produce life, a biofriendly universe need not necessarily produce culture and highly aware valuing agents. But because a biofriendly universe worldview would imply the ubiquity of life throughout the universe, advocates would also acknowledge that many instances of culture could also arise. Such instances might even be likely given the number of chances for life to arise throughout the universe, the variety of phenomena natural selection can produce over time, and the usefulness of intelligence and culture for the fitness of individuals and groups (Dawkins 1986; Dennett 1995).

However, in a biofriendly universe, the ubiquity of intelligence and culture, while noteworthy as a prevalent physical phenomenon, would not necessarily carry with it any particular cosmic significance. Culture would be seen merely as a derivative of biological systems (or any agents with sufficient interests to create "culture") that are consequences of selection pressures and natural laws—none of which has any particular significance, direction, or broader purpose for the universe as a whole. Indeed, while "simpler" forms life may originate fairly easily and frequently throughout the universe, some views suggest that complex life and the origin of intelligence and culture could be so highly historically contingent as to be exceedingly rare (Gould 1989; Ward and Brownlee 2000).

6.3 The Role of Intelligence, Value, and Culture

Our speculations become more interesting as we continue along the spectrum shown in Fig. 6.1. A weaker "bootstrapped universe" view would suggest that (a) the universe has in some sense "bootstrapped" itself into the realm of value (e.g., including meaning, purpose, ethics, aesthetics, etc.) via the emergence of life, intelligence, and culture; but that (b) there is no broader significance of that bootstrapping for the universe as a whole beyond merely having those interesting properties emerge as a product of cosmic and biological evolution. The implications for such properties in the universe are primarily confined to cultural beings and perhaps valuing agents more generally, but would not constitute any particular cosmic significance (and hence is shown on the "one-way relationship" side of the spectrum.)

The kind of cosmological reverence explored here, particularly as a foundation for a cosmological theory of value, can prompt us to more specifically consider matters of normative aspiration in light of cosmic evolution and cosmological reverence—including biological evolution by natural selection as a key dynamic when thinking about normative aspiration of humans or other intelligent beings (a subject touched on previously). We will first briefly reflect on normative aspiration in light of natural selection and then in light of cosmic evolution and cosmological reverence—with a particular emphasis on how such considerations may relate to putative intelligent beings in the universe more broadly. We will then briefly touch on additional pragmatic considerations regarding the search for, preparation for, and potential interactions with extraterrestrial life and intelligence.

6.4 Normative Aspiration in Light of Natural Selection

As noted previously, the explanation of biological systems evolving by natural selection, in its simplest form, tells us that the evolution of life and intelligence have been governed essentially by what can be thought of as "filtered replication." Filtered replication can be seen as a very general process by which entities that are copied either continue copying or stop copying based on factors that prevent their continued replication. If filtered replication is a reasonable way to think about biological evolution, then that may help point to the possibility that sufficiently aware beings—e.g., beings aware of their evolution being shaped by natural selection (or "filtered replication")—may eventually view such a process as inadequate or "undesirable" in at least two ways: (1) It could be seen as an inadequate "design process," and (2) it could be seen as something that is not sufficiently meaningful or aspirational for "highly evolved" beings.

6.4.1 Shortcomings of Natural Selection

As noted above, one way that filtered replication may be seen as inadequate is as a design process in general. While natural selection has clearly produced a wide variety of successful forms of life and has ultimately led to human intelligence and human culture, it also took a very long time and has arguably been very destructive, blind, painful, and cruel by many standards. Also, from a general design perspective and from the perspective of specific design metrics, there is much about biological systems today that appear to be bad designs that humanity has spent much time trying to fix—as the history of medical pursuits shows (including mental health which has many ties to evolutionary psychology).[1]

Secondly, and perhaps more importantly, a species aware of the primary motivation of natural selection to serve the sole purpose of enhancing replicating effectiveness may choose to fundamentally question the goal and aspirational value of biological replication or replication more generally—especially as a motivator for highly intelligent beings. Natural selection may be seen by advanced beings as a simple, crude, and narrowly single-minded process—a process that just happens as a matter of replicative dynamics and nothing more.

The complex products resulting from the simplicity of natural selection makes it profoundly intellectually appealing and is arguably the most important idea in human history (Dennett 1995). But the philosophical and normative implications may be very unappealing to many intelligent beings. On both means and ends, evolution by natural selection may be subject to harsh scrutiny by sufficiently aware beings, perhaps to the point of trying to explicitly transcend it or overturn it—as much human culture has arguably already attempted. However, it may also be that intelligent beings will decide not to devalue natural selection and the powerful engine of self-interest. Such a choice (captured at a high level in Fig. 3.1) may depend critically on the psychology and rational explorations of such beings—much of which human cultural evolution has experimented with, intentionally and unintentionally, successfully and unsuccessfully (Sowell 1987; Haidt 2012; Cockell 2015).

[1] However, filtered replication or "genetic algorithms" can be a powerful tool for finding effective solutions depending on the fitness functions used, the kind of solution sought, and the kinds of tools performing the computation (Hillis 1991; Lupisella 2004).

6.4.2 Discovering Natural Selection: A Turning Point?

Normative aspiration in light of natural selection may be a key turning point for intelligent beings. It may lead to strong intentional pursuits to move beyond much, if not all of the programming of biological evolution. Super intelligences will presumably overcome the physical frailties of biology, and may also be able to deal very effectively with the pitfalls of self-interest. This could end up simply being an intellectual philosophical choice for sufficiently advanced beings who may have much control over what can be realized. At the very least, for advanced beings, normative aspiration in light of natural selection may be seen as a potential source of intense philosophical exploration and experimentation.

We may need to allow for the possibility that self-interest is so deeply embedded in self-replicating systems that it ultimately cannot be changed in any fundamental way. Although attempts to control self-interest are important to human life, such attempts, especially in the extreme, have also had unfortunate consequences throughout human history. Such attempts may be fundamentally untenable and even immoral—they are often sources of tension at the root of fundamental ideological conflicts that are manifested in endless domestic political disputes and international relations. Selfishness may essentially keep beings in something like a "selfishness trap" that most or all beings ultimately do not change—either because they are not able to or because they choose not to. The very nature of selfishness may make it impossible for selfish beings to significantly moderate, modify, or give it up—even if it were relatively easy and many were willing to attempt it (Lupisella 2001).

Truly transcending selfishness may be futile if it is an "all-or-nothing" requirement in the sense that all beings must fully participate in transcending their own self-interest, otherwise "cheaters" could destabilize and undermine whatever fairness, cooperation, and stability exists (Trivers 1971; Axelrod and Hamilton 1981; Axelrod 1984/2006, 1997). It may also be that because self-interest can reap so many benefits for so many individuals and societies that intelligent beings consciously, or less consciously, choose to at least allow, if not encourage, high degrees of self-interested behavior. Recognizing that self-replicating entities may forever be bound to self-interest can significantly inform how we think about ourselves and extraterrestrial beings and if or how we might interact with extraterrestrial beings.[2]

6.4.3 Beings Without Natural Selection?

It may be that some extraterrestrial beings, or "unitary" solitary beings, did not evolve via natural selection and without social selection pressures (if, for example a designer was involved or if there are other ways a single being might exist), in which case, normative aspiration in light of natural selection would not likely be

[2] Seth Shostak, in *Confessions of an Alien Hunter* (2009a), considers the possibility that machine intelligence will still evolve via competitive forces.

relevant. But even in scenarios of this kind, normative aspiration could still be a preoccupation of such beings, and perhaps more importantly could be an important part of a general shared framework to help us understand and interact with other beings, including a very foreign unitary solitary being (Lupisella 2013). Regardless, an understanding (and perhaps reverence) of cosmic evolution and the myriad details of how it appears to have given rise to life and intelligence would presumably shape philosophical deliberations of such beings.

6.4.4 An Infinite Possibility Space?

A unique challenge regarding normative aspiration is that the possibility space for ideal preferred futures is arguably infinite. Navigating such a space, and doing so with other potential extraterrestrial beings, especially those that did not evolve via natural selection, raises profound challenges for how and what advanced beings may value, what they may aspire to. In one extreme, it may turn out that highly capable life-forms without any form of normative aspiration, or very different kinds of normative aspiration, could be very dangerous (Michaud 2007). In another extreme, knowledge of cosmic evolution and the dependence of life and intelligence on that evolutionary process can prompt a broader sense of value and perhaps even reverence for the universe and all within it.

6.5 Normative Aspiration in Light of Cosmic Evolution and Cosmological Reverence

As touched on previously, modern cosmology has accumulated a substantial body of evidence and compelling theoretical frameworks to suggest our universe has physically evolved over approximately 13.8 billion years. The universe has expanded and cooled, forming structures such as galaxies, stars, and planets, for which at least one has biological systems that have led to self-aware beings with knowledge of cosmic evolution. Our high level of cosmological awareness can inform, and be directly part of, our normative aspiration and may already be part of the normative aspiration of other intelligent beings elsewhere. The cosmos can be seen as a commonly shared, and perhaps commonly understood, frame of reference for beings throughout the universe.

6.5.1 Unidirectional Cosmic Relationships

Unidirectional or "one-way" formulations of cosmic-cultural relationships suggest that the cosmos is significant for culture in that culture arises from cosmic evolution and the cosmos informs and influences culture in critical ways. Cultural evolution, or valuing agents more generally, however, have no particular significance for the universe at large in these kinds of "unidirectional" perspectives such as cosmological reverence. And while such unidirectional formulations may be uninteresting and trivially true to some, such views may have interesting implications nonetheless. At a minimum, there may be an implication that one of the great challenges for intelligent cultural beings may be to learn to cope with, and perhaps finally accept, a profound and deep sense of uncertainty regarding any larger cosmic sense of meaning and purpose. Such uncertainty may have to be treated as a kind of empirical question to be possibly addressed over very long time periods as evidence is accumulated—but perhaps without ever obtaining a satisfactory answer. Coping with the uncertainty of broader cosmic objective meaning may be one of the most profound challenges sufficiently aware beings have to face, and this could have profound implications for cultural evolution—as it arguably already has with human beings.[3]

In addition to the uncertainty of broader cosmic significance, it may be that intelligent beings might have to learn to cope with a perceived likelihood of cosmic *insignificance*, leading some to a kind of nihilistic worldview. It may be that species move from uncertainty about broader objective cosmic meaning and purpose to high levels of confidence that there is indeed no such thing for intelligent beings. For others, something short of nihilism might suggest instead a kind of "cosmically local" relativism where value, meaning, purpose, ethics, and aesthetics derive solely from the affairs of cultural beings who think, behave, and ultimately freely choose in such ways as to establish widely accepted norms and standards to help "local" beings coexist and collectively seek parochial values.

Even if a single instance of intelligence and the associated emergence of cultural evolution were to eventually spread throughout the universe, unidirectional cosmic-cultural views would still suggest there is ultimately no particularly deep cosmic significance for the emergence and long-term implications of cultural evolution. However, in the same way that Jungian archetypes may be thought to reflect deeper collective realities of human experience and possibly deeper realities more generally, many separate instances of cultural beings throughout the cosmos independently coming to similar conceptions of value may also imply a deeper cosmic

[3]Despite the prevalence of broad worldviews (e.g., religions) and the apparent angst that a large part of the human population exhibits regarding broader meaning and purpose in their lives, many people and cultures seem to be more comfortable than is commonly thought in coping with the uncertainty they face about any larger "cosmic" meaning or purpose (Lupisella 2019). Although, a Pew Research Center report, "The Global Religious Landscape" (2012), notes that a wide range of unaffiliated people appear to still hold religious or spiritual beliefs such as believing in God or a "higher power," e.g., as low as 7% of Chinese adults and as high as 68% of adults in the United States.

reality and significance for cultural evolution. But such a perspective would still suggest that appearances of "cosmic-cultural convergence" are not necessarily reflective of deeper cosmic realities—they may instead merely reflect the realities and implications of biological and cultural selective processes and nothing more. Claims that such cosmic cultural convergence would reflect a deeper cosmic reality would fit more into bidirectional cosmic-cultural worldviews which will be explored shortly.

Perceptions of a lack of objective meaning and purpose of the universe may also prompt intelligent beings to choose to see themselves in some sense at odds with a hostile universe and perhaps at odds with all other creatures that are potential competitors. Indeed, this appears to be a legitimate "realist" scientific and practical way to see our biological condition in a cosmic context since most of space the wider universe appears to be inhospitable to our form of life. Further, the second law of thermodynamics appears to generally work against biological systems over time. Much of what we do, even within the comfort of our terrestrial cocoon, is targeted at controlling and changing our environment to make it more suitable to our basic biological and psychological needs which are arguably ill-suited for long-term survival in an unfriendly universe. And to date, we have no evidence of intelligent life elsewhere, which can be interpreted to suggest our origins may be rare and random (Ward and Brownlee 2000).

6.5.2 Epistemic, Scientific, and Value Dimensions of Cosmological Reverence

The brand of cosmological reverence intended here can be seen to have at least three major components: epistemological, scientific, and philosophical—all of which are somewhat interrelated. As explored previously, epistemological relationalism seems to at least be implied by the modern physics of relativity theory and quantum theory. Many, if not most, interpretations of what those theories imply about how we know the world suggest that it is only knowable if we acknowledge the relationship of the observer and observed that observations and measurement results are ultimately relative. In the important case of relational quantum mechanics, all that can be said about the world is how a quantum state relates to some other state. What might this mean for a theory of value, particularly in the context of cosmic evolution? Our epistemic dependence on relations can be interpreted to imply a certain amount of caution and humility regarding the ways in which we think we understand the world and our attempts to control it. Further, what may ultimately be most valuable are relationships, interactions, and process, in contrast to the objects and "things" of philosophical materialism.

The science of cosmic evolution tells us that we are intimately bound up with and critically related to the cosmos at least in the sense that we are a product of a long and complex cosmological evolutionary process. The recognition of our cosmological origins, along with the emerging myriad of scientific details of cosmic evolution, can imply that we can in some important sense revere the universe, its overall evolution, and perhaps cosmological processes in the most general sense. Hence, we, or intelligent beings more generally, can choose to incorporate the universe as a whole into worldviews, and then into pursuits of normative aspiration since we may see the physical universe as the ultimate totality of reality, an ultimate entity, an ultimate source, or at least a "ultimate" frame of reference. Intelligent beings may see the universe as an unfolding story in which they and many kinds of beings have various kinds of significance because they are related to the universe as a whole—with an emphasis on being a product of cosmic evolution. If beings come to understand the universe as an *evolving* cosmos, recognizing their evolution from it in a deep sense, this can give rise to a universe as a "creator" of sorts, or perhaps at least an ultimate source of creative activity.

As Fig. 1.1 shows, cosmological reverence is a weaker form of relationalism and hence a weaker basis for a cosmological theory of value in the sense that it is primarily a recognition of our intimate relationship to our universe through the lens of contemporary science and nothing more. Cosmological reverence does not require any significant philosophical leap other than to plow through the naturalistic fallacy and make the philosophical choice to value our universe and cosmic evolution based on what science has taught us about the universe and our relationships to it. This is a theory of value that is relatively non-prescriptive, very open-ended, and realized primarily by valuing agents that are sufficiently self-aware and have the ability to learn about their universe, contemplate what it may mean, and make the choice to see the universe and its evolution as valuable.

Consistent with an open-ended interpretation of this kind of value, this minimal value acknowledges the universe as our source, as part of our "big history" (Christian 2004)—including allowing for the recognition that the universe is not exactly made for us today in the sense that it is not consistent with our physical health if we leave the cocoon of our biosphere without technology to protect us. So cosmological reverence does not require a blind reverence or blind commitment to cosmic evolution or the universe as a whole—especially at the expense of other interests and commitments we may have. The "softness" in cosmological reverence as suggested here is a feature of sorts. We can still choose to note and value our broader relationship to the universe and potentially derive forms of value and ethical positions from that, while at the same time avoiding problematic implications when prioritizing values and developing ethical positions. The universe need not be the dominant center of ethical commitments in cosmological reverence worldviews.

Related to the idea of a softer version of cosmological reverence, Sean Carroll writes:

> We talk about 'awe and wonder,' but those are two different words. I am in awe of the universe, its scope, its complexity, its depth, its meticulous precision. But my primary feeling is wonder. Awe has connotations of reverence: 'this fills me with awe and I am not worthy.' Wonder has connotations of curiosity: 'This fills with me with wonder and I am going to figure it out'. I will take wonder over awe every day.

Carroll understandably puts distance between awe and wonder, but it may be more than is warranted. Awe can indeed connote reverence, and in the context of this chapter and broader book, that is a reasonable and useful implication.[4] But reverence can come from both awe and wonder. And importantly, reverence need not imply any relative lack of worthiness. Reverence, while sometimes implying relative degrees of worthiness in some contexts, can also simply imply deep respect and nothing more.

But the difference between awe and wonder is still useful. It may be a predisposition or aesthetic leaning to emphasize one over the other, or it may be a careful rational choice.[5] And while both can imply reverence for the universe, wonder is arguably more practically motivating with its curiosity and boldness to understand our universe—to apprehend that with which we are in awe. Such wonder, and its implications for comprehending and appreciating our universe, can help serve as a bridge from cosmological reverence to cosmocultural evolution that we will explore in Chap. 7.

Stronger versions of cosmological reverence can make the universe a compelling priority in a worldview. Such views could be consistent with what could be considered weaker forms of "cosmocentrism" (as shown in Fig. 6.1, overlapping with cosmological reverence). Cosmocentrism, in the context of cosmological reverence, makes the universe a high priority based primarily on the knowledge of our universe as a critical source of our origins and possibly "sustenance." Cosmocentrism is a general notion that need not ascribe spiritual or divine significance to the universe, but can still nevertheless be interpreted to see the universe as an ultimate source of meaning, purpose, and value—hence making it a priority in a worldview. Cosmocentrism can then help point to the possibilities for "cosmocentric ethics" (as shown in Fig. 1.1) that may ultimately be able to provide guidance for activity of intelligent beings in the universe—stronger versions of which will be explored in subsequent sections.

6.6 Additional SETI Considerations in Light of Cosmological Reverence

Here we will briefly explore some additional potential near-term and long-term practical implications of cosmological reverence that we might consider when thinking about how to prepare for the potential discovery of extraterrestrial life and intelligence, including how we may guide future research and perhaps help inform policy measures and initiatives that relate to preparing for such discoveries. These considerations are not mutually exclusive, but in some cases they have opposing implications. The suggestions are intended to help explore a series of considerations that may help

---

[4] When I had sent a letter of condolence to Ann Druyan, Carl Sagan's widow, noting how sometimes the "awe weighs heavily," and having in mind how Carl used to talk of the positive power of awe, she responded with the penetrating confirmation: "yes, the awe weighs heavily, but it also uplifts."

[5] I am reminded of the thoughts by a main character, Nicole Wakefield, in the *Rendezvous with Rama* series, written by Arthur C. Clarke and Gentry Lee: "*The whole key to life is understanding*" and "*I will not be afraid because I understand.*" Perhaps most notably, "*And understanding is happiness.*"

in preparing for a discovery of extraterrestrial life where the details of putative beings and the state of humanity are highly uncertain, but where some form of importance or value is placed on the relationships of intelligent beings to the cosmos as a whole.

6.6.1 Cosmic Humility and Rational Analysis

An essential concern when engaged in the kind of speculation we are exploring regarding beings for which we have no data, particularly speculation that involves values and normative aspiration, is that we should exercise a high degree of humility and open-mindedness. We simply do not know the modalities of other putative beings. They may exist in a very large possibility space of values that are extremely foreign and potentially incomprehensible to us, or perhaps the process of natural selection tends to shape similar outcomes. While modern cosmology has taught us much about our universe and our place in it, it has also revealed further questions and mysteries.[6] Our knowledge of the universe is presumably still very limited, and so we should consider that our understanding of natural selection and cosmic evolution, while important to the considerations in this chapter, may be very incomplete and even irrelevant to other beings, particularly if we inhabit a multiverse that may have different physical realities. The way in which other beings may view and value the universe could be quite different from how we might value it.

The history of human values and human ethical aspiration can certainly be interpreted to suggest we cannot rely on the existence of objective answers to guide our normative aspirations or those of other beings, perhaps suggesting extreme humility in our exploration of what is valuable and how to pursue it. Considering that much of human history has been ruled by ethical certitude, often grounded in religious beliefs, a potentially helpful thought experiment may be to consider scenarios where other intelligent beings claim to have certitude about very different sets of norms and value theory more generally—some of which may be tied to forms of cosmic reverence. How would we cope with this kind of extreme diverse certainty or incomprehensible "dogma"?

Nevertheless, having noted the need for humility, our knowledge of natural selection, cultural diversity, and cosmic evolution may make attempts to explore possible motivations and values of other intelligent beings a reasonable and worthwhile pursuit—a pursuit possibly within reach (Cohen and Stewart 2002). It will presumably be helpful to think carefully and diversely about potential long-term futures for humanity and life on Earth, which can also inform how other intelligent beings may exist and behave throughout the universe and what the impact of their discovery and our interactions with them may be. There are many ways this kind of analysis can proceed (Dick 2015, 2018), including applying various methods to our own rich history here on Earth (Dick 2015a).

---

[6]For example, our universe's expansion may be accelerating, suggesting the presence of another force at work in cosmic expansion (Riess et al. 1998; Perlmutter 1999).

6.6.2 Extraterrestrial Normative Aspirations and Expectations

If normative aspiration is a preoccupation of other beings in the universe, and particularly if normative aspiration turns out to be an important dynamic in cosmic evolution overall, then intelligent beings may at least desire, or possibly even expect, continually improving normative aspiration from all intelligent cultural species. Advanced beings may go further and expect a certain level of normative aspiration to be reached before civilizations are allowed to be part of a galactic club, or a better, a cosmic club! If other beings aspire to see themselves in a cosmic context, and to act consistent with such a worldview, this may be something for humanity to be sensitive to when considering what other beings might value, what they might desire, and how that might affect our interactions. As noted earlier, such worldviews may also be worth pursuing as a way to have a possible common ground of value with other beings.

As an example, many products of cosmic evolution could be seen to have value and deserving of a certain level of ethical commitment from a rational normatively aspiring species. Such considerations may inform our ethical posture toward non-intelligent life and non-living objects more generally, either here on Earth or in off-Earth environments (Rolston 1990, McKay 1990, Smith 2009, 2014, Schwartz 2020, Smith and Mariscal 2020). Perhaps some degree of ethical commitment to non-intelligent life, such as putative microbes on other planets, might be part of this kind of respect for other intelligent beings that may value that life in a way we may not. Such commitments might be part of the extraterrestrial normative expectations noted previously (Lupisella and Logsdon 1997).

6.6.3 Message Interpretation and Active SETI

Among the more likely extraterrestrial interaction scenarios to consider is that a signal may be received from a distance great enough that there will not be any immediate or direct physical interactive consequences. Nevertheless, understanding the true meaning of a message could be critically informed by general and specific matters of normative aspiration and cosmic perspectives such as cosmological reverence. The universe is something we presumably have in common with other beings in our universe and could serve as a shared frame of reference for understanding and communicating with each other in general as well as about what is valued and why.

Humanity may also seek out extraterrestrial intelligence by proactively sending signals where we think intelligence might exist (often called "active SETI" or also METI (messaging extraterrestrial intelligence). A more thorough exploration of the implications of normative aspiration in light of natural selection and cosmic evolution may inform if, how, and when we might wish to engage in active SETI and METI, and perhaps more importantly, what we might wish to say—including views that explicitly express a certain kind of value or reverence for our universe (Vakoch 2009).

6.6.4 Machine Intelligence and Artificial Morality

In a post-biological universe where intelligent beings are machines, there may also be aspirations for "artificial morality" as well. We may want to go well beyond something like Isaac Asimov's laws of robotics to help protect humans and other non-machine beings (e.g., see Gardner 2009 for a related treatment). We may wish to contemplate a much broader set of moral considerations for other beings and perhaps the universe as a whole. Indeed, it may very well be easier for machines in a post-biological universe to be more objectively ethically committed because they may not have the deep genetic motivations of self-replication and self-interest. For example, they may be programmed, or self-programmed, for complete fairness. They may decide on their own that what they aspire to is true fairness and equity—and it may turn out not to be too difficult as the cost of caring for others goes down as beings advance (e.g., due to technical advancements and social and organizational constructs), causing the overall "caring capacity" of advanced societies to go up (Lupisella 2013).

Machine intelligence, or superintelligences more generally, may be able to choose among many diverse and unusual values to guide and realize their actions. Artificial intelligence could have a kind of "artificial reverence" for the universe given that the universe could be recognized as an ultimate "source." As we explored briefly before, superintelligences may even wish to create universes to help contribute to the diversity of a multiverse and the creation of novel life-forms and intelligence (similar to the themes explored by Olaf Stapledon noted previously).

6.6.5 Diversity

Many of the above considerations relate to how we and other intelligent beings view diversity, how it is valued, and how beings cope with it. Cosmic evolution and biological evolution have produced a stunning array of life and mind in what appears to be a highly open-ended manner, suggesting the need for much more comparative psychology, including non-human animals (Marino 2015). Values of beings throughout the universe could be unimaginably diverse, suggesting the need for careful contemplation and many thought experiments informed by a high degree of intellectual humility and respect. Preparing to deal with such a large possibility space of diverse values might be important to try, but it might also turn out to be futile. Regardless, there is a large amount of diversity right here on Earth, including non-human animals that exhibit remarkably diverse behaviors. Understanding them at much deeper levels could be very informative to preparing for discoveries of other forms of life and intelligence in the universe.

A better understanding of human diversity and human rationality and irrationality, partly informed by evolutionary psychology, social psychology, cognitive science, and neuroscience, may help us prepare ourselves for dealing with the discovery of extraterrestrial intelligence and may also help us better understand potential psychologies of non-human intelligences. Cautiously applied, analogs in the social and behavioral sciences may be just as useful as in the natural sciences. Given what is at stake, it seems reasonable to continue to carefully explore the diverse prospects for extraterrestrial life and intelligence in the cosmos.

6.6.6 A Post-Intelligent Moral Universe

Dick (2003) has suggested the “Intelligence Principle” as a central force of cultural evolution, whereby intelligence and knowledge will be the primary motivators of most advanced species. However, it may be that knowledge could reach a point of diminishing returns in some broader philosophical context. That is, it may be that advanced cultures become so capable that intelligence and knowledge will become less important because what may matter most is what advanced beings value, since what they value will likely be the realities they create. Highly capable and aware beings may come to see intelligence merely a means for many open-ended value pursuits and goals—first as ends sought by selfish replicators that they may wish to appreciate, but that they may also recognize as too limited. Increasingly unconstrained goals emerging from diverse desires and infinite imaginations of valuing agents may be the ultimate drivers of highly advanced beings and the realities they create. Such beings may desire and pursue something more akin to a *morally creative cosmos*, a universe in which the central pursuits are not necessarily intelligence, knowledge, and technology, but fairness, caring, and diversity (Lupisella 2009, 2013).

In a “post-intelligent” universe, a guiding principle may be something more like a “values principle” or “wisdom principle” where what also matters, and what may increasingly matter most, is what beings value and why—perhaps critically tied in some way to revering the universe as a whole. The pursuit of intelligence and knowledge, along with normative aspiration, can be a complimentary dance and lead ultimately to what could be profound forms of wisdom. This fact–value interplay represents a key human endeavor and is arguably a critical pursuit for the human species, particularly as we become a more capable and truly globally integrated space-faring species. Careful steps in this kind of cosmic waltz between facts and values, between cosmos and culture, can help us prepare for potential future discoveries of extraterrestrial life and intelligence, and for a future in which intelligent beings play significant roles in cosmic evolution as whole—something we will explore further in the next chapter on cosmocultural evolution.

6.7 Summary of Chapter 6: Cosmological Reverence

In this chapter, we explored *cosmological reverence* as the first of three cosmological worldviews that draw from contemporary science and a relationalist framework. Cosmological reverence is a relatively "weak" relational view in the sense that it emphasizes our fairly well understood scientific relationships to the universe and ascribes a certain value to those relationships, but does not suggest that intelligent beings have any significance for the universe as a whole. In that sense, cosmological reverence is a kind of one-way or unidirectional cosmological worldview as shown in Fig. 6.1.

Cosmological reverence has varying degrees and implications—partly based on the prospects for life in a cosmic context where a biofriendly universe and increasing trends toward complexity can be interpreted as stronger versions of cosmological reverence. We explored the idea of normative aspiration in the contexts of both natural selection and cosmic evolution and what it might mean philosophically and more pragmatically for considerations regarding extraterrestrial life and intelligence, including the basic assumption that we presumably have the universe in common with other beings in our universe. We recognized the possibility that long-lived intelligent beings, potentially extremely powerful beings, may see their normative aspirations as the ultimate drivers of the wisdom they seek and the realities they create, which can inform how we think about ourselves and how we might interact with other beings in the universe.

References

Axelrod, R., & Hamilton, W. D. (1981). The evolution of cooperation. *Science, 211,* 1390–1396.

Axelrod, R. (1984, 2006). *The evolution of cooperation* (Rev. ed.). Perseus Books Group.

Axelrod, R. (1997). *The complexity of cooperation: Agent-based models of competition and collaboration*, Princeton University Press.

Carroll, S. (2016). *The Big Picture: On The Origins of Life, Meaning, and Universe Itself.* New York: Dutton.

Christian, D. (2004). *Maps of time: An introduction to big history*. Berkeley: University of California Press.

Cockell, C. (Ed.). (2015). *The meaning of liberty beyond earth*. New York: Springer.

Cohen, J., & Stewart, I. (2002). *Evolving the alien*. London: Ebury Press/Random House.

Dawkins, R. (1986). *The blind watchmaker: Why the evidence of evolution reveals a universe without design*. New York: W. W. Norton & Company Inc.

Dennett, D. (1995). *Darwin's dangerous idea*. London: Penguin Books.

Dick, S. J. (1996). *The biological universe: The twentieth century extraterrestrial life debate and the limits of science*. Cambridge: Cambridge University Press.

Dick, S. J. (2000). Cosmotheology: Theological implications of the new universe. In S. J. Dick (Ed.), *Many worlds: The new universe, extraterrestrial life, and the theological implications* (pp. 191–210). Philadelphia: Templeton Foundation Press.

Dick, S. J. (2003). Cultural evolution, the postbiological universe and SETI. *International Journal of Astrobiology, 2*(1), 65–74.

Dick, S. J. (Ed.). (2015). *The impact of discovering life beyond earth*. Cambridge University Press.

Dick, S. J. (2015a). History, discovery, analogy: Three approaches to the impact of discovering life beyond earth, In S. J Dick (Ed.) *The impact of discovering life beyond earth* (pp. 38–54). Cambridge University Press.
Dick, S. J. (2018). *Astrobiology, discovery, and societal impact.* Cambridge University Press.
Dick, S. J., & Strick, J. (2004). *The living universe: NASA and the development of astrobiology.* Piscataway, NJ: Rutgers University Press.
Gardner, J. (2009). In S. J. Dick & M. Lupisella (Eds.), *The intelligent universe. cosmos and culture: Cultural evolution in a cosmic context.* Washington, D.C.: NASA History Series. Gould, S. J. 1989. *Wonderful Life: The Burgess Shale and the Nature of History*. New York: W. W. Norton & Company.
Gould, S. J. (1989). *Wonderful life: the burgess shale and the nature of history*. New York: W. W. Norton & Company.
Haidt, J. (2012). *The righteous mind.* New York: Random.
Hillis, D. (1991). Co-evolving parasites improve simulated evolution as an optimization procedure. In C. Langton, C. Taylor, J. Farmer, & S. Rasmussen (Eds.), *Artificial life II, (Santa Fe Institute Studies in the Sciences of Complexity, Proceedings, 10)* (pp. 313–324). Addison-Wesley: Redwood City, CA.
Lupisella, M. (2001). Participant statement. In *Humanity 3000, Seminar No. 3 Proceedings, Bellevue, WA: Foundation for the Future* (pp. 37–38 and also discussed on p. 251, 330). http://www.futurefoundation.org/documents/hum_pro_sem3.pdf.
Lupisella, M. (2004). Using artificial life to assess the typicality of terrestrial life. *Advances in Space Research, 33,* 1318–1324.
Lupisella, M. (2009). Cosmocultural evolution: The coevolution of cosmos and culture and the creation of cosmic value. In S. J. Dick & M. L. Lupisella (Eds.) *Cosmos and culture: Cultural evolution in a cosmic context* (pp. 321–359). NASA SP-2009–4802. http://history.nasa.gov/SP-4802.pdf.
Lupisella, M. (2013). Caring capacity and cosmocultural evolution: Potential mechanisms for advanced altruism. In D. A. Vakoch (Ed.), *Extraterrestrial altruism*. Berlin: Springer.
Lupisella, M. (2019). Is the universe enough? Can it suffice as a basis for worldviews? *Journal of Big History, 3*(3), 123–140.
Lupisella, M., & Logsdon, J. (1997). Do we need a cosmocentric ethic? In: *Paper IAA-97-IAA.9.2.09 Presented at the International Astronautical Federation Congress*. Turin: American Institute of Aeronautics and Astronautics.
Marino, L. (2015). The landscape of intelligence. In S. L. Dick (Ed.), *The impact of discovering life beyond earth*, Cambridge University Press.
McKay, C. (1990). Does mars have rights? In D. MacNiven (Ed.), *Moral expertise*. London: Routledge.
Michaud, M. A. G. (2007). *Contact with alien civilizations: Our hopes and fears about encountering extraterrestrials*. New York: Springer.
Palmeri, J. (2009). Bringing cosmos to culture: Harlow shapley and the uses of cosmic evolution. In S. J. Dick & M. Lupisella (Eds.), *Cosmos and culture: Cultural evolution in a cosmic context.* Washington, D.C.: NASA History Series.
Perlmutter, S., et al. Measurements of omega and lambda from 42 high redshift supernovae. *Astrophysical Journal, 517*(2), 565–586. arXiv:astro-ph/9812133.
Pew Research Center Report. (2012). The global religious landscape: A report on the size and distribution of the world's major religious groups as of 2010. https://assets.pewresearch.org/wp-content/uploads/sites/11/2014/01/global-religion-full.pdf.
Riess, A. G., et al. (Supernova Search Team) (1998). Observational evidence from supernovae for an accelerating universe and a cosmological constant. *Astronomical Journal, 116*(3), 1009–1038. arXiv:astro-ph/9805201. Bibcode:1998AJ...116.1009R. https://doi.org/10.1086/300499.
Rolston, H. (1990). The preservation of natural value in the solar system. In E. C. Hargrove (Ed.), *Beyond spaceship earth: Environmental ethics and the solar system*. San Francisco, CA: Sierra Club Books.

Schwartz, J. S. J. (2020). *The Value of Science in Space Exploration*. New York: Oxford University Press.
Shostak, S. (2009a). *Confessions of an alien hunter: A scientist's search for extraterrestrial intelligence*. Washington D.C.: National Geographic.
Smith, K. C. (2009). The trouble with intrinsic value: A primer for astrobiology. In C. Bertka (Ed.), *Exploring the origin, extent, and future of life: philosophical, ethical and theological perspectives* (pp. 261–280). Cambridge: Cambridge University Press.
Smith, C. K. (2014). Manifest complexity: A foundational ethic for astrobiology? *Space Policy, 30*, 209–214.
Smith, K. C. & Mariscal, C., Eds. (2020). *Social and Conceptual Issues in Astrobiology*. New York: Oxford University Press.
Sowell, T. (1987, rev. in 2007). *A conflict of visions: Ideological origins of political struggles*. New York: Basic Books.
Swimme, B., & Berry, T. (1992). *The universe story: From the primordial flaring forth to the Ecozoic era: A celebration of the unfolding of the cosmos*. Harper.
Trivers, R. L. (1971). The evolution of reciprocal altruism. *Quarterly Review of Biology, 46*, 35–57.
Vakoch, D. (2009). Encoding our origins: Communicating the evolutionary epic in interstellar messages. In S. J. Dick & M. Lupisella (Eds.), *Cosmos and culture: Cultural evolution in a cosmic context*. Washington, D.C.: NASA History Series.
Ward, P. D., & Brownlee, D. (2000). *Rare earth: Why complex life is uncommon in the universe*. New York: Copernicus Books.

Chapter 7
Cosmocultural Evolution

7.1 Background and Overview

In this chapter, we move left to right in Figs. 1.1 and 6.1 from cosmological reverence to a stronger "bidirectional" relationship between cosmos and culture, namely *cosmocultural evolution,* which that emphasizes the coevolution of cosmos and culture in which cosmos and culture are important for each other. In this chapter, we will explore some practical near-term and longer-term implications, including a long-term worldview that can be characterized as a "morally creative cultural cosmos"—a post-intelligent, post-technological universe that enters the realm of conscious evolution driven largely by moral and creative pursuits of valuing cultural agents.[1]

We should avoid such a strong distinction between cosmic evolution and cultural evolution that they are thought of as entirely separate. Cultural evolution is ultimately a part of cosmic evolution in the broad sense that culture appears to have emerged as part of the physical evolution of the universe. But we can still make a useful distinction to the extent that it can help address the interesting question of how significant cultural evolution may be in a cosmic context.

[1]For a treatment of "conscious evolution," see Banathy's (2000) *Guided Evolution of Society: A Systems View*, and Hubbard's (1998) *Conscious Evolution: Awakening the Power of Our Social Potential*. New World Library.

[2]See Dawkins (1976) for an introduction to the notion of "memes" that are suggested to be a kind of cultural replicating unit (e.g., an idea, song, social norm, etc.), and Blackmore (1999) for an expanded treatment.

Much of this chapter is taken from a previous book chapter from Lupisella (2009a).

M. Lupisella, *Cosmological Theories of Value*, Space and Society,
https://doi.org/10.1007/978-3-030-25339-4_7

Culture has helped life on Earth, particularly *Homo sapiens*, survive and thrive in ways that sometimes defies belief. What human beings have created, what we have become and are becoming, is remarkable, inspiring, and sometimes shocking. But is it an illusion of sorts? Is culture merely an increasingly complex result of biologically driven self-interest, arising from the happenstance of life? Is it merely a blind walk, or run, of replicating memes—the cultural equivalent of natural selection?[2]

While it may be true that much, if not all, culture could ultimately be explained directly and indirectly by Darwinian explanations, it may also be true that cultural evolution is beginning to break free of our biological heritage. Natural selection has been working on the experiment of life for close to 4 billion years on Earth, and what we witness now with human culture is so rich, so complex, and so uncertain that we have to wonder how it will evolve and how it may be evolving elsewhere in the universe.

Other species on Earth arguably exhibit basic forms of culture, but those instances appear to be far less complex, and perhaps far less meaningful than what human beings experience. Our technology, art, and what we know of our world, is unspeakably exhilarating and terrifyingly dangerous. We are capable of powerful creations and complete annihilation. Our consciousness is uncontainable to the point of agonizing and uplifting awareness (including an awareness of our awareness!). Modern humans appear to have an usual power to say the least.

To some, this anthropocentric cheerleading may seem like the worst kind of "speciesism"—a kind of blind, unethical delusion engendered by biologically driven affinities for one's own likeness. But exaltation of humanity in no way justifies unchecked devotion at the expense of others who inhabit our Earth and potentially worlds beyond. Nevertheless, the evidence seems clear: human beings are running away with culture. And culture may be running away with us.

7.2 Bootstrapped Cosmocultural Evolution

> We're only here for a short while. And I think it's such a lucky accident, having been born, that we're almost obliged to pay attention. We are — as far as we know — the only part of the universe that's self-conscious. We could even be the universe's form of consciousness. We might have come along so that the universe could look at itself. [3]
>
> —Mark Strand, Poet Laureate

Indeed, we are looking at the universe. But we are doing more. The notion of a "bootstrapped universe" (touched on previously and shown in Fig. 6.1) that bootstraps itself into the realm of value via cultural beings such as ourselves, however random its emergence may be, need not be confined to a one-way relationship such

[3]Interview with Mark Strand in: Csikszentmihalyi (1996). Maria Popova nicely explores Strand's quote and related sentiments in the "Artist's Task to Bear Witness to the Universe" on her Brain Pickings site: https://www.brainpickings.org/2015/01/28/mark-strand-creativity/.

as cosmological reverence where culture and value do not have any particular significance for the universe. We can claim for ourselves, and the universe at large, that we are not just a way for the universe to look at itself (as the poet Mark Strand suggests above), but to *value* itself. We can entertain that paying attention and manifesting value in the universe via valuing cultural agents such as ourselves has noteworthy significance for the universe, in part because the universe now has potentially important properties it might not have had otherwise (e.g., value, meaning, purpose), and also because the potential of valuing cultural agents for cosmic evolution may be highly significant and influential—perhaps even unlimited.

Cultural evolution having importance for cosmic evolution might also be called the "cosmocultural principle", suggesting a sufficiently different kind of evolution is emerging—the coevolution of cosmos and culture, where culture plays an important and perhaps ultimately critical role in the overall evolution of the universe. Cosmos and culture can coevolve and highly influence each other in this kind of worldview. Strong versions of cosmocultural evolution could be interpreted to suggest that not only is cultural evolution in some sense "on par" with physical cosmic evolution, but also that cultural evolution might have unlimited potential in cosmic evolution and may ultimately be more important than the "physical" cosmic evolution that gave rise to cultural evolution.

As shown in Fig. 6.1, teleological, pantheistic, and theistic worldviews can be interpreted as very strong versions of cosmocultural evolution, in part because in those views cultural evolution can be thought to be inherent in the universe for reasons that will be touched on in the next section. *Bootstrapped cosmocultural evolution*, however, suggests that while culture is an important aspect of the universe and its future evolution, it is not important for any particular additional reason other than culture simply arose via physical and biological evolution and now has the power to be significant for the universe. This is a universe that has bootstrapped itself into the realm of value, meaning, and purpose via its own otherwise non-teleological, or non-purposeful physical evolution. Cultural evolution was not baked in. It just happened.

Yet the universe may nevertheless be entering a qualitatively different and perhaps fundamentally new and unlimited kind of evolution via the emergence of cultural beings. This is a worldview that (a) makes few assumptions about the nature of the universe, while (b) advocating that the universe has bootstrapped itself into the realm of value, meaning, and purpose, and (c) allows for the possibility of significance and possibly even unlimited potential for cultural evolution in the universe, where *strong* bootstrapped cosmocultural evolution would suggest the possibility for unlimited influence of culture on cosmic evolution (as shown in Fig. 6.1).

While bootstrapped cosmocultural evolution is consistent with the broader notion of "cosmocentrism", one could still advocate for a bootstrapped cosmocultural evolutionary view without necessarily making the universe the central priority or source of value. One could believe that it is significant that the universe has bootstrapped itself into the realm of value via cultural agents such as ourselves and still also maintain an anthropocentric or "ratio-centric" (Smith 2009) worldview in which intelligent beings are still the ultimate priority.

The notion of bootstrapped cosmocultural evolution relies on minimal claims about the nature of the universe in the sense that it stresses the observation that "properties" such as value, meaning, purpose, and culture have appeared in the universe as a result of the appearance of replicating molecules (which may have emerged essentially by chance) and results of natural selection operating on those replicating systems over long periods of time leading to intelligence and culture. The appearance of culture (e.g., mechanisms of transferring knowledge, social stability, artistic pursuits, etc.) would not necessarily be so surprising in this worldview. Many things can happen by chance, including those that might be considered low probability.

What might be more surprising than the emergence of culture is the origin of life itself. The emergence of replicators appears to work in opposition to the second law of thermodynamics, at least on "local" scales, creating local spatial-temporal negentropy. While this "negentropic" dynamic is still ultimately consistent with the second law in the sense that biological systems are essentially open systems that increase the entropy of the universe overall, biological evolution is still arguably a quite different kind of evolution than the rest of cosmic evolution that has come prior. However, progress has been made in understanding replicating systems and there appear to be plausible explanations that could account for the origin of molecular replicators (Fry 2000). Indeed, claims are often made, if not implied, that the universe is teeming with life. Many scientists point out that the biofriendliness of the universe's "fine-tuning" of physical laws and constants is compelling and needs to be explained (Barrow and Tipler 1986; Davies 2007). However, one reading of the evidence is that the universe is not so biofriendly.

It may turn out that the universe is teeming with life and perhaps intelligent life. But it seems premature to jump to such a conclusion in the absence of sufficient evidence. As Carl Sagan often cautioned, absence of evidence is not evidence of absence, so we must simply do the experiment and keep searching—perhaps for a very long time. Indeed, an obvious response to the concern that we have not yet found life is that we have only begun the search, and in such a large universe, it may take a very long time to find life. But the Fermi paradox is less easy to dismiss—despite many creative responses (Webb 2002): if life is ubiquitous and intelligence and technology follow, then, as Fermi asked, "Where are they?" Intelligent life, or signs thereof, should arguably be easier to find than more primitive life—if they are not hiding. Perhaps the nature of the universe lends itself to producing life (e.g., replicating systems), but not necessarily to producing complex life and the intelligence and culture that has resulted on Earth (Ward and Brownlee 2000).

But if mere replication is the key, we can imagine that it could happen under many diverse physical circumstances. We can also imagine that once replication is underway in a dynamic environment, the emergence of increasingly diverse and perhaps eventually highly complex strategies for replication (e.g., including sociality and culture) would not be so surprising given enough time and given the very simple and compelling mechanism that is natural selection.

7.3 Teleology, Pantheism, Theism

As we move along the spectrum of Fig. 6.1, we can also briefly touch on other broader categories of worldviews that can be interpreted to place cultural evolution in a cosmic context. Teleology, pantheism, and theism are all influential concepts and worldviews that have implications for how cultural evolution may be seen in relation to our universe.

Teleology has traditionally been used to suggest design, purpose, or "final causes" in both biology and cosmology and has often been associated (although not exclusively) with supernatural explanations. There are many ways to think about teleology, and the term has understandably fallen out of favor among many scientists for a variety of reasons (Mayr 1992; Davies 2007[4]), but still receives attention from scientists and philosophers (Manson 2003). For the purposes of this chapter and the book more broadly, we can highlight a relatively weak form of teleology that is more akin to suggesting the existence of cosmological trends or natural directionality that could be consistent with loosely defined notions of "progress" or "purpose". This is to be distinguished from pantheism and theism that contain the additional features of divinity and transcendence, respectively. This weaker form of teleology has been implicitly or explicitly suggested by a number of scientists ranging from (a) suggestions of trends toward increasing self-organization and complexity (Kauffman 1995; Chaisson 2005), to (b) life and intelligence as "cosmic imperatives" or inevitable cosmic phenomena (De Duve 1995; Lloyd 2006; Davies 2007), to (c) "multiverse" and/ or "anthropic" worldviews that suggest our particular universe is made for life (Smolin 1997[5]; Rees 1997; Carr 2007), to (d) more explicit eschatological treatments (Teilhard De Chardin 1955; Tipler 1994) that have stronger pantheistic and theistic aspects.

Pantheism generally equates God with the universe and tends to reject the notion of a personal and/or transcendent God (although some religions—especially Eastern traditions—tend to be pantheistic while retaining a "personal" nature for God).[6] Unlike most of the teleological views noted above, pantheism is closer to a theistic and religious position where sacredness and divinity play important roles (MacIntyre 1967; Levine 1994; Harrison 1999). There are many conceptions of pantheism,

---

[4]In his book, *Cosmic Jackpot* (2007), Paul Davies devotes several sections to teleology and indicates its controversial nature by titling the first of those sections, "Tackling the T-Word" (p. 233).

[5]Lee Smolin's proposal is interesting as it relates to this weak form of teleology because he suggests a cosmic selection mechanism (often referred to as "cosmological natural selection" which is somewhat analogous to biological natural selection) that essentially "selects" or filters for universe's like ours (or at least universes that have characteristics for being relatively stable and long-lived). This arguably allows for an interpretation that our universe arises from a somewhat directional selection mechanism (e.g., for "stable" long-lived universes), that once selected, may have directionality toward life and possibly intelligence, and then toward giving "birth" to subsequently similar universes.

[6]Taoism, certain forms of Buddhism and Hinduism, and some mystical strands of monotheism have pantheistic features.

ranging from Eastern religions such as Taoism, to mystical versions of Western religions, to purely naturalistic views based on biology and cosmology that focus on the realities of our natural world and the universe at large (Dick 2000).[7]

The significance of culture in pantheistic worldviews could be interpreted a number of ways. Many thinkers who revere the universe might be dissatisfied with implications of cultural evolution that result in the blind consumption and destruction of our natural world.[8] However, much of culture could be viewed with reverence since culture is a part of, and product of, the sacred and divine whole that is the universe. By including culture as an emergent part of the universe, pantheism can be seen to imply noteworthy significance for culture in a cosmic context. But culture may also be seen in such a way as to be ultimately undifferentiated from all other phenomena in the universe, suggesting that culture is not necessarily more significant than other "divine" manifestations of cosmic evolution. Other pantheistic views may ascribe more significance to culture by claiming that cultural evolution is a way for the universe to become self-aware, to in some sense know itself, and perhaps ultimately to become more beautiful, more divine. Such interpretations would certainly imply that cultural evolution occupies a special place in cosmic evolution.

Theism generally claims the existence of a transcendent, personal, supernatural God who is omnipotent, omniscient, creator of and active in our universe, and generally immutable.[9] However, there are noteworthy exceptions to almost all of these characteristics, of which immutability is of particular relevance for this book. Theistic views that promote the idea of mutability tend to incorporate evolutionary concepts and cosmic evolution to understand the nature of God. Process theology (e.g., Cousins 1971; Cobb and Griffin 1976; Epperly 2011), "evolutionary theism"[10] (Haught 2000), and what might be called "eschatological cosmic evolution"—which envisions an essentially theistic "God" primarily at the end of cosmic evolution which results from the evolution of intelligence within the universe (e.g., Teilhard De Chardin 1955; Tipler 1994)—are all forms of theistic thought that involve deep evolutionary processes. The role of cultural evolution in such worldviews is arguably strong, at least in the sense that intelligent beings and their behavior are often thought

[7] As noted previously in Chap. 2, on precedent, Dick (2000) emphasizes the idea of "cosmotheology" and the prospects for a purely "natural God" to help better incorporate the physical realities of cosmic evolution in theological worldviews.

[8] Brian Swimme is an example of many writers with backgrounds in cosmology who express deep reverence for the universe and our natural world based on views of physical cosmological evolution (*The Universe Story* 1992 and *The Hidden Heart of the Cosmos* 1995), but he is critical about human impacts on the Earth's environment.

[9] "Deism" is distinguished from theism in that deism tends to see God as not being active in or "interfering" with the world. Panentheism (with the added 'en' in the middle) sees God as imbued and active within the world as part of the nature of the universe (as in pantheism), but also as transcendent, essentially making the universe a subset of a larger God. Deism and panentheism are considered subsets of theism for the purposes of this treatment, in part because they advocate the important distinguishing feature of transcendence.

[10] "Theistic evolution" or "evolutionary creationism" can be seen as a more narrow pursuit that attempts primarily to integrate biological evolution with traditional religion.

to reflect and/or in some way be directly or indirectly connected to "God". Such theistic views would imply a significant role for cultural beings in the cosmos as robust manifestations of the processes of cosmic evolution and/or perhaps as "co-creators" of reality.

7.4 The Complexity and Power of Human Culture

While the origination of basic culture may not be surprising given replication and natural selection acting over long timescales, what is noteworthy is the level of sophistication of human culture, the depth of our awareness, and the extent and speed that culture has evolved and influenced an entire planet. What human beings are doing with culture, what culture is doing with us, and what culture is doing to our world and beginning to do to worlds beyond, are all noteworthy to say the least. The apparently unlimited potential of cultural evolution calls out for our attention. But it is not at all clear whether the level of sophistication we see with human culture should somehow be an expected outcome of cosmic or biological evolution. Nevertheless, culture has emerged. It is here, and it is powerful.

Irrespective of whether the emergence of life, intelligence, culture, and value is a low probability, it need only happen once. Given enough time and space, surprising things are bound to happen. It may also be that the universe possesses value completely independent of valuing agents—a worldview that will be explored later (Rolston 1990; Lupisella 2009b, 2016). But what we can claim with confidence today, what we know about the universe now, is that the cosmos has the properties of value, meaning, purpose, and culture through us. The universe literally has those properties within it, within us, where it otherwise might not have without the emergence of valuing cultural beings such as ourselves (and possibly other forms of life that have similar characteristics).

So even with this "minimalist" bootstrapped cosmocultural perspective, we can assert that the universe has now become a different kind of entity—an entity that contains culture, manifesting value to extreme degrees. Those qualities are at least in our evolutionarily driven predispositions, in our interests, in our culture, in our worldviews—and hence in the universe. Regardless of origin and form, value is indisputably manifested in the universe through cultural beings like ourselves. What is not so obvious is how significant that really is.

7.5 Limited Ontological Significance

The use of "ontological" in what follows is meant partly in the traditional philosophical sense of having to do with that which exists (physical or non-physical), or with "being." However, it is also intended here to emphasize a narrower sense than that broader use sometimes implies—namely "physical being" or "physical existence," with the caveat that "physical" is still intended in a fairly broad and somewhat loose sense, implying, among other things, that it is not necessarily limited to our formulations of physical reality as we understand it today.[11]

The emergence in the universe of properties such as value and meaning is noteworthy in that the universe has produced something different and has perhaps become something quite different to the extent that it has evolved in an interesting and important way by creating value, meaning, purpose, and culture. This may be seen as a form of limited ontological significance in that (a) emergent properties such as value are primarily physical manifestations within valuing agents and cultural beings who have evolved with sufficiently complex value-laden interests, and (b) while the universe may be changing in an important yet limited physical sense via the emergence of value, the broader significance beyond that is minimal or negligible. Nevertheless, culture is significantly ontologically relevant for the cosmos as a vehicle of that change—what may be thought of as a kind of limited physical change, yet perhaps a qualitative change as well. That is, a new kind of qualitatively different physical manifestation has emerged—namely *value*, along with culture as a way of further manifesting and operationalizing value. We may think of this limited ontological significance as corresponding to a weak form of bootstrapped cosmocultural evolution.

We, and potentially other beings, may see ourselves as a means by which the universe has been bootstrapped into the realm of value, culture, meaning, and purpose. Sufficiently aware beings may bring to the universe not only normative aspiration, but in that pursuit of normative aspiration, such beings may also explicitly choose to value the universe and in some sense act on its behalf. Through our perceived roles as creators and arbiters of value, intelligent cultural beings may act as a kind of cerebral cortex for universe. We may all see ourselves as effectively making the whole of the universe valuable merely by developing these kinds of worldviews (and perhaps acting on them). So as suggested previously, intelligent cultural beings may not just be a way for the universe to know itself, but to also *value* itself. This is a different kind of anthropic principle, maybe a kind of "anthropic value principle": the universe is valuable because we are here to value it.

As sources and arbiters of value, cultural beings not only have the ability to recognize this ontological significance, but in some sense can deepen it by emphasizing it. Simply choosing to adopt and emphasize this ontological significance for ourselves and for the universe makes that significance more significant. It can become more

[11]This use of "ontological" that stresses a broader sense of physical existence is arguably consistent with the definitions of some philosophers. For example, Mario Bunge defines ontology as "…the serious secular version of metaphysics. The branch of philosophy that studies the most pervasive features of reality, such as real existence, change, time, chance, mind, and life. Ontology does not study constructs, i.e., ideas in themselves." He goes on to write: "General ontology studies all existents, whereas each special Ontology studies one genus of thing or process – physical, chemical, biological, social, etc. Thus, whereas general ontology studies the concepts of space, time, and event, the ontology of the social investigates such general sociological concepts as those of social system, social structure, and social change. Whether general or special, ontology can be cultivated in either of two manners: speculative or scientific" (Bunge 1999, pp 200–201).

deeply meaningful to human beings in part because as sources of value, *we* confer that deepening upon the cosmos. This limited ontological significance, however, is akin to having a kind of "benign" worldview that does not have much, if any, practical consequence, but which might nevertheless offer helpful and/or comforting worldviews and perhaps ultimately form a foundation for more practical implications.

7.6 Foundations for a Cosmocultural Evolutionary Theory of Value

If cultural evolution has significance or roles in cosmic evolution, either now or perhaps more likely in the future, then value can be seen to be shifted slightly away from physical cosmic evolution and more toward cultural beings and their evolution. On stronger versions, this can be interpreted to suggest that beings may have some degree of obligation or duty to their wider universe and its evolution. In that sense, cosmocultural evolution is a little more about our future relationship to the universe and future choices we make to advance the interests of beings within a larger cosmic web as part of a broader balancing act with the wider universe and all other beings that may co-inhabit the cosmos with us.

This kind of view arguably lies somewhere between the extremes of teleology and non-teleology (although perhaps closer to the non-teleological end) in the sense that if there has been no meaning, purpose, or direction prior in the universe, there is now in the form of highly intentional and highly aware valuing agents which may more strongly create cosmic evolutionary trends in the future. The universe may not have ever been intrinsically teleological, but it may be bootstrapping itself into the realm of teleology in the sense that cultural valuing beings can bring a kind of direction to the universe if they choose—through an emerging process of cosmocultural evolution that can give rise to some kinds of trends or directionality. So, in a sense, the universe may be bootstrapping itself into something teleological. Some kind of directionality, and perhaps even purpose, may now be possible and perhaps already underway due to beings like ourselves that have the capacity for purposeful pursuits. The trends or directionality could presumably be very broad, diverse, and ever-changing, and any specific outcome would not be predetermined. In bootstrapped cosmocultural evolution, no one specific trend would necessarily be intrinsic to the fundamental nature of the physical universe. Instead, any directionality or trends that may arise would emerge via the complex and dynamic intentionality of deliberate valuing agents and nothing more.

7.7 Practical Cultural Significance

A lack of external objective meaning for human life may be unsatisfying to many, caught forever in endless cycles of relativism, a morass of unbearable responsibility for our own meaning and purpose, and perhaps ultimately for that of the universe. But it looks like choice is inescapable.[12] And while choice can sometimes be weighty and even debilitating, it can also be liberating and empowering—so much so that freedom and responsibility form critical pillars of most modern human societies.

What then are we to do with the destructive and constructive power of value-based culture and our freedom to realize it in so many ways? What kind of "cosmic" beings should we strive to be? Or, perhaps a more immediate challenge: Why should we worry about such long-term questions at all? Why should we contemplate our role in the universe when it seems so distant, so far into the future, so uncertain? Such considerations may not be as distant as first glances might suggest. There are a number of relatively near-term practical challenges that could have consequences for human behavior that relate to these broader longer-term cosmic perspectives, among them: globalization, biospheric stewardship, space ecology, search for extraterrestrial intelligence, off-Earth migration, and long-term survival and development.

7.7.1 Globalization: Transcending Tribalism

While "localism" is an important and often healthy counterbalance to the forces of globalization, it is important to strike balances between the two. Knowing how our evolutionary heritage can blind us to longer-term implications, and more specifically, how it can drive blind group identity (Haidt 2012), can help us be more careful about such proclivities. Seeing ourselves as having a role in a cosmic context that suggests our selfish biological evolution is not necessarily part of an inevitable deep purposeful cosmic design can help motivate us to take better control of our local and collective global behavior as a species. It can help sensitize us to some of the blinding adverse effects of cultural forces such as dogmatic ideologies that too often lead to unnecessary conflict.

Seeing ourselves has having significance in a longer-term cosmic context can help us envision a healthier, more united human species, creating recognition of value for global engagement and collective global pursuits as opposed to pursuing strictly group or national interests. Seeing ourselves as a special fragile species that may be "on our own," yet with potential cosmic significance, can help us act as a

[12]Even if free will may be a kind of illusion in a "deterministic" universe, the way most of us experience and act in the world, individually and collectively, is through intentional choices with consequences. Robert Wright writes: "History, even if its basic direction is set, can proceed at massive, wrenching human cost. Or it can proceed more smoothly—with costs, to be sure, but with more tolerable costs. It is the destiny of our species—and this time I mean the inescapable destiny, not just the high likelihood—to choose" (Wright 2001, 10).

global species on behalf of each other and our environment. And the need to come together better as a species is evident on many fronts—some of which is noted in what follows.

7.7.2 Biospheric Stewardship

While most people today would acknowledge the importance of preserving and improving our environment, certainly at least for human survival and quality of life, it is somewhat surprising that we are not better stewards of our biosphere. Or is it? As noted above, biological evolution does not quite program us to be sensitive to longer-term, broader, non-obvious implications—and perhaps for good reason from a "selfish gene" perspective. Higher probability near-term consequences would be expected to heavily influence animal behavior given how natural selection is driven by what appears to be essentially blind genetic replication into the next generation.

Indeed, despite our awareness of long timescales, extinction, and our own power to potentially mitigate catastrophiesc and extinction-level threats, it is noteworthy that humans are still primarily reactive near-term creatures—especially in many Western cultures. But it is changing. What we have learned about our planet, about our biosphere—much of which is a result of human instinct for looking up and out into the cosmos—has led us to see ourselves as a "pale blue dot" in a vast cosmic ocean.[13] The way in which we appreciate and deal with that fragility, the way we have begun to see ourselves as biospheric stewards—perhaps increasingly in the context of cosmic evolution—has and should continue to influence how we care for our world and how we value life here on Earth or elsewhere in the universe.

7.7.3 Space Ecology

Humanity is beginning to have a direct impact on our space environment—however small that impact may be for now. We have created much debris in low-Earth orbit, including remains from weapons shooting down satellites. We have crashed, landed, lived, and played on the Moon.[14] We have sent robotic spacecraft near and far. We have leaked signals and intentionally sent communications into deep space. We are living in near-Earth orbit on an International Space Station. And, we are now planning to permanently live on the Moon, Mars, and beyond.

---

[13] A NASA Voyager image, looking back at our solar system showing Earth as a very small light blue dot "suspended" in a sunbeam, inspired Sagan's 1994 book, *A Pale Blue Dot*, in which he wrote eloquently about Earth and humanity occupying such a small part of a vast cosmos.

[14] Alan Shepard famously "played" golf on the Moon, see: http://www.youtube.com/watch?v=AdqBL5pdRT8.

These developments raise a range of environmental questions, ranging from if and how we should preserve certain space environments (Almar 2002; Williamson 2006), to how we can be more responsible, eco-friendly explorers (Cockell 2007), to how we might deal with the prospects of extraterrestrial life (Lupisella 1999; Dick 2015, 2018). How we deal with such questions will be informed, in part, by our own cultural evolution, by how we see ourselves in the context of our wider universe, and by what we see as our responsibilities within broader contexts and worldviews. Space agencies around the world take substantial measures to avoid the contamination of certain space environments such as Mars, but it is primarily for scientific reasons. What about other perspectives, including broader philosophical and ethical perspectives, which might inform such policies? (See COSPAR Workshop Report on Ethical Considerations for Planetary Protection 2012). A broader cosmic perspective can help inform such considerations, and cosmocultural evolution implies a much broader set of responsibilities that could inform those considerations.

7.7.4 The Search for Extraterrestrial Intelligence

As touched on in the previous chapter, our longer-term cosmic considerations can inform how we think about other intelligent beings in the universe, and if/how they may communicate and act throughout the cosmos. Considerations having to do with forms of cosmocultural evolution can inform if and how we communicate with, and perhaps ultimately interact with, other intelligent beings. We have already intentionally and unintentionally sent communications into outer space. It is unlikely that our transmissions have been detected by other civilizations for a number of reasons, but nevertheless, our communications are leaking out into space, and on occasion, being sent intentionally.[15]

Perhaps more importantly, there have been serious treatments of a more systematic attempt to send communications from Earth to potential extraterrestrial civilizations (Vakoch 2004, 2008, 2009). What would we communicate in such attempts? How would we decide what to say? Our considerations about cultural evolution in a cosmic context, our own specific cosmic perspectives, and the daunting plurality of our views, will presumably inform such decisions, if not be explicitly articulated in great detail in communications with extraterrestrial beings (Lupisella 2011). If cultural beings have significance for the universe, we might view seeking and interacting with other cultural beings as part of a larger cosmic pursuit—a pursuit to seek out, share, and perhaps co-create value, meaning, and purpose for the cosmos through interactions of many cultural species.

---

[15]For example, Messaging Extraterrestrial Intelligence (METI) International transmitted a signal in late 2017 in the direction of GJ 273, known also as "Luyten's Star," 12 light-years away. There have been a number of intentional transmissions over the past several decades, including a transmission from Arecibo Radio Telescope in 1974 that is claimed to be detectable almost anywhere in the galaxy.

7.7.5 *Off-Earth Migration*

While the preservation and enhancement of planet Earth can be a central organizing priority for humanity, we could also benefit from attempting migration beyond our home world. Humanity will almost certainly benefit from doing the difficult experiment of migrating off-Earth to assess if and how we can effectively and sustainably survive and thrive outside the comforts of our natural biosphere. If we do not, we may run a higher risk of extinction (Shostak 2009). It is this kind of "experimental ethos" and experimental migration narrative that can be a significant justification for space activities.[16]

And there are other benefits to migrating off-Earth. Human beings have slowly, and perhaps sometimes too painfully, benefited from social experimentation that has often been driven and accelerated by migrating to new environments with new challenges and new freedoms (Cockell 2015). Experimenting with new forms of social organization and new means of governance can benefit from the challenges of new environments. Migrating into the wider universe can serve that purpose and help unite nations in a common long-term endeavor of human growth and social experimentation.

But there is also a less practical, and perhaps equally important consequence of migrating off-Earth: the creation of cultural diversity. Finding better ways to live is important. But finding *different* ways to live is interesting and culturally enriching to the human experience (and perhaps to the "non-human" experience as well). New branches of cultural evolution can enhance the human condition and enrich human existence by giving us more to take note of, more to study, more to choose from, more to appreciate, more to participate in, more to take joy in, more to be inspired by, more to be in awe of, more to love.

Cultural diversity, and perhaps diversity in general, may have practical benefits (i.e., having a wide variety to choose from as circumstances require),[17] but diversity may be a value in its own right, an end unto itself and worth pursuing for its own sake (something that will be explored further in subsequent chapters).[18] Given the potential for quite diverse life-forms throughout the universe, diversity may have broad cosmic significance beyond our own aesthetic appreciation. And so our motivations for extraterrestrial migration, and the associated new branches of human cultural evolution, can be informed by broader long-term cosmic perspectives that can incorporate diversity as a key philosophical motivator.

---

[16]We should also bear in mind cautions against overzealous and perhaps ill-informed justifications of space activities that may make misguided assumptions such as the universal appeal of exploration or assuming space settlement is required to save humanity (Schwartz 2017).

[17]In Global Brain (2000), Howard Bloom stresses the importance of "diversity generators."

[18]Personal communication with Jill Tarter, October 1997. Freeman Dyson has noted that diversity is a "great gift" and that its fostering is a "great goal" (Dyson 1988).

7.7.6 Long-Term Survival and Development

The above relatively near-term issues have long-term trajectories, potentially of cosmic significance. Why not treat them that way? Why not take a truly long view, a cosmic view? If we look long-term, what might we see? What visions might we pursue? Let us have the courage to face the uncertainties of such contemplations. Let us not shrink in the face of complex threats to our species, or be passive about presently unknown threats. Let us seek out and mitigate such threats as no other species has been able to do. Let us rally our political institutions and global resources to become a truly long-lived species (Tough 1991; Lupisella et al. 2003; Smith 2008).

Now is the time to be proactive about our long-term survival and development—whether by protecting our planet from asteroid impacts and gamma-ray bursts or by migrating off-Earth to reduce our chance of extinction and to create new branches of human civilization. Understanding ourselves in a cosmic context, seeing ourselves as potentially significant for the universe, can help motivate and inform such endeavors, including, perhaps ultimately—as fanciful as it may sound—surviving the "end" of the universe.

7.8 A Cosmocultural Evolutionary Scale: From Cosmic Demotion to Cosmic Promotion?

We do not know of course whether there will be, or are presently, deep or large-scale physical or non-physical consequences of culture for the universe at large, but a cosmocultural perspective suggests it is a plausible enough possibility. This is where the unlimited potential of cosmocultural evolution becomes more interesting, maybe even compelling. Cultural evolution is ultimately manifested as behaviors at what are often large-scale social actions, so if we envision cultural evolution acting over long timescales, especially cosmic timescales, we can imagine potential impacts for the universe as a whole.

In 1963, Nikolai Kardashev proposed three types of civilizations categorized by the amount of energy that can harnessed: a Type I civilization harnesses roughly the energy associated with a planet; Type II, a star; Type III, a galaxy. Carl Sagan calculated Earth to be something like a .7 civilization (not quite Type I) and further suggested the additional categorizing criterion of information available to the civilization (Sagan 1973). Vidal (2014) speculates that some binary systems may be indicative of Type II civilizations feeding on stars—what he calls "starivores." Ćirković (2004) suggests that Type IV should be used to designate a civilization that can harness the power of its supercluster. Kaku (2005) suggests a Type IV civilization could harness extragalactic energy sources such as dark energy, and Galantai (2004) has suggested a Type IV level which harnesses the energy of the visible universe.

Here, we will explore a somewhat different scale that overlaps with what is been proposed prior, but has a different emphasis and extends our thinking further in what is admittedly more speculative and more qualitative than the energy scales noted above. The scale we explore here emphasizes the kind of impact and influence a culture exerts on its environment and the universe, which is related to, but not limited by the energy it is capable of controlling.

7.8.1 Type I Influence: Planetary

Planetary influence would include the ability to influence a planet and solar system bodies (e.g., biospheric control, defense from astronomical impacts such as asteroids, etc.). Humanity is close to this kind of influence now in some respects, but far in other respects—i.e., despite our global climate impacts, we are probably far from effective planetary climate control. We should also consider the possibility that life may not necessarily originate or evolve on planetary bodies. For example, life may originate or evolve in interstellar clouds or possibly even in "free space"—perhaps near a star or other astrophysical energy sources. This would suggest a more general category title such as "localized influence" where the environmental influence and control of the species is confined to a relatively local scale—e.g., solar system type scale.

7.8.2 Type II Influence: Astrophysical

Given the possibility that advanced intelligences may have the ability to modify large-scale structures in the universe, ranging from stars and solar systems to galaxies and black holes, it may be fruitful to consider how particular kinds of aspiration might result in specific kinds of astrophysical modifications or "engineering." Astrophysical influence would imply a capacity for using, controlling and modifying astrophysical objects on small and large scales—e.g., stars and galaxies, superclusters, possibly black holes, etc. Examples would be the ability to harness most, if not all, of a star's energy (such as Dyson Sphere), control the energy output of a star, extend the lifetime of stars, modify the composition of stars, control the energy of galaxies and superclusters, possibly create black holes, and harness unusual forms of energy such as "dark energy."

Deutsch (1997) has suggested that knowledge is a kind of fundamental physical quantity, and as an example, he uses the intentional modification of stellar evolution (to prolong the lifetime of a planet's sun) as a way to illustrate how intelligent beings might use their knowledge to alter large-scale cosmic phenomenon and as a result affect the "knowledge" of observers of that star when they observe that it does not fit their standard models. This, and even more physically transformative examples, would be cases where knowledge could transform physical reality on astrophysical scales.

7.8.3 Type III Influence: Cosmological

If advanced beings are able to create black holes, they might use them as energy sources or to create universes—as some have begun to contemplate (Guth 1994; Smart 2009). Cosmological influence implies an ability to influence and control phenomenon on cosmological scales, i.e., the large-scale dynamics of the universe, but within the constraints of physical laws and constants. Examples might be extending the lifetime of the universe (perhaps by slowing or accelerating expansions or contractions), possibly transmitting something like information through a big crunch, creating baby universes, or creating an information processing universe and/or a kind of cosmic mind. We might also take seriously the possibility that we are living in an artificial or simulated universe and that universes might be regularly simulated and/or created merely as part of a creative endeavor by superintelligences?

7.8.4 Type IV Influence: Ontological

As noted previously, "ontological" is perhaps used slightly differently here than some traditional uses in that the ontological influence suggested here applies an ability to control and modify the physical nature of the universe itself—truly "mind over matter." As an example, this would amount to an ability to change physical constants and perhaps laws. This might apply to the selfish biocosm hypothesis proffered by Gardner (2003, 2009), and could also be consistent with views that suggest life and mind are key creative drivers of a "self-synthesizing" and/or participatory universe (Davies 2009; Wheeler 1988, 1989) whereby observers participate in shaping all of physical reality, including laws, particularly as mind and cosmos eventually merge (Davies 2009).

7.8.5 Type V Consequence: Metaphysical

At the risk of treading into even deeper waters, but for the sake of completeness, we can consider the possibility of another category, one that is perhaps more a matter of consequence than influence (although "influence" would not necessarily be ruled out)—namely, metaphysical. Here too "metaphysical" is used in a somewhat non-traditional philosophical sense. It is intended primarily to capture that which may be considered to be beyond physical reality—an often popular use of the word—to the understandable chagrin of many scientists and philosophers.

One way to think about metaphysics in this context is that while ontology is concerned with what actually exists (primarily physically), metaphysics is more concerned with what may exist, or theoretical possibilities, including that which is "non-physical," based on a general fundamental understanding of reality in the broadest sense. This may include considerations such as God, or a kind of Platonic realm in which there are theoretical properties (e.g., "universals" such as mathematical constructs, logic, redness, etc. (Armstrong 1989)), or theoretical possibilities for the universe and reality in general. The domain of metaphysics might be thought of as the largest possibility space for "ultimate reality,"[19] a subset of which is the actual and/or temporary state of reality. On some interpretations, this metaphysical possibility space could include things like value, meaning, purpose, divinity, "spirit," etc. So, metaphysics can be seen to be a very broad (perhaps the broadest) category of investigation (and experience).

If one thinks of value and cultural evolution as somehow transcending physical reality, if even only partially,[20] then cultural beings may be seen as partly metaphysical beings and are arguably beginning to have metaphysical significance for the universe by manifesting value for the universe. Whether a bootstrapped cosmocultural perspective can be interpreted to go so far as to imply metaphysical significance is highly speculative and problematic of course—not the least because of the speculative nature of metaphysical considerations in general. But if there is any metaphysical significance to consider, some interpretations of bootstrapped cosmocultural evolution could be consistent with suggesting there may be partial metaphysical relevance for cultural evolution to the extent that emergent phenomena (ultimately rooted in, but perhaps not fully constrained by physical reality—e.g., things like value and cultural evolution) sufficiently "transcend" physical reality nonetheless.[21] However, bootstrapped cosmocultural evolution would in no way be committed to such a view, and is arguably more consistent with no such transcendence because bootstrapped cosmocultural evolution emphasizes that value and cultural evolution is bootstrapped from the physical universe we see and does not require an appeal to "non-physical" reality.

7.8.6 *Further Reflections on Scale*

In the context of this proposed cosmocultural evolutionary scale, one way of interpreting bootstrapped cosmocultural evolution (especially stronger versions that emphasize unlimited potential) is that we are beginning to have planetary (or "local") influence, we are studying for astrophysical influence, we are contemplating cosmological influence, we are speculating about ontological influence, and we may

---

[19] van Inwagen (1998) suggests that metaphysics is an attempt to sufficiently generally describe "ultimate reality." Alston (1998) also examines the notion of ultimate reality and considers the relevance of "possibilities" with respect to the notion of ultimate reality.

[20] The notion of "dualism" in the philosophy of mind claims that in some respects, mental phenomena are non-physical (Hart 1996). As will be discussed more later, it may even be that something like love is, or can become, an important force in the universe that is in some sense, "metaphysical."

[21] However, it is not at all clear if/how "emergence" from physical systems gives rise to anything that truly transcends physical reality—what might be called a kind of "non-physical emergence."

have a kind of metaphysical consequence if value and cultural evolution somehow transcend physical reality.

These levels do not have to represent "progress" per se. It may be that a deeper wisdom reflects contentment with a species' planet (or whatever smaller locale it may exist in) and their ability to live sustainably without expanding. A species' local existence could arguably still have significance or consequence. For example, the choice to exist with a smaller physical impact on the universe may allow for greater diversity, more opportunities for other life-forms, and a kind of deep contentment that might be an example for other intelligent beings to follow. Being content and fulfilled with a local reality can provide a balance to evolutionary tendencies to seek out more, to consume more, to become more.

7.8.7 A Cosmic Promotion?

Scientists and thinkers have long been fond of pointing out humanity's "great demotions." From Copernicus to modern-day cosmology (perhaps with the exception of "anthropic principles" and associated observations of "fine-tuning"), humanity has been displaced and demoted from privileged positions in the cosmos. Perhaps it is time for a promotion—one that goes beyond the confusion of anthropic principles and does not rely on teleological assumptions and assertions about the ultimate nature of the universe. Bootstrapped cosmocultural evolution allows for the possibility that life, intelligence, and culture could have arisen by chance, while at the same time suggesting that such phenomena are cosmically significant. Stronger versions suggest that cultural evolution may have unlimited significance for the cosmos—physically and philosophically. Our means of origin and cosmic location should not be confused with our cosmic potential.

As valuing agents, we can claim and create a more meaningful universe. For some, this may mean the creation of, or at least recognition of, a more evolved, more significant, more complex and diverse universe. This may strike some as anthropocentrically superficial, but in the view of cosmocultural evolution the value we humans bring to the universe is at least noteworthy. In the same way that intelligence and culture are impacting planet Earth, we may also ultimately have so much freedom and power as to impact the universe itself. And in the same way we seek to strike balances between individual freedom, collective well-being, and our environment, we may ultimately also wish to seek such balances for beings everywhere and for the whole of the universe as we become an increasingly cosmic species.

7.9 A Morally Creative Cultural Cosmos: A Busy Utopia?

> But equally it appeared to us as unreasoning Creativity, at once blind and subtle, tender and cruel, caring only to spawn and spawn the infinite variety of beings, conceiving here and there among a thousand inanities a fragile loveliness.

—Olaf Stapledon, *Star Maker*

With great potential, comes great responsibility. So what do we do with the potential of cultural evolution and our values that drive it? Culture can have insidious negative effects of course—a kind of "culture curse"—especially on non-human life and the environment. As we increasingly wrap ourselves in culture, our basic humanity, our common humanity, our connection to each other and our broader environment—especially the global environment—is often masked, if not completely lost. Indeed, human beings can lose themselves in culture. But culture can also uplift and inspire. Culture has resulted in large-scale devastation as well as magnificent human achievement. A critical challenge we face is coping with the dramatic variances for what is thought to be justified destruction and laudable human activity.

How much can we control cultural evolution anyway? Susan Blackmore warns of what might be called "runaway memetic evolution," whereby replicating memes blindly go about their replicating business—unchecked and unfettered—resulting in massive, often unforeseen destruction (Blackmore 2009). But it does appear that human beings can indeed control cultural evolution to some extent, perhaps to an extent that implies we should take responsibility for it. After all, we are certainly in large part, the creators of culture.

As noted previously, Dick (2003) has proffered the "intelligence principle," for which he writes: "The maintenance, improvement and perpetuation of knowledge and intelligence is the central driving force of cultural evolution, and that to the extent intelligence can be improved, it will be improved." He goes on to say: "The intelligence principle implies that, given the opportunity to increase intelligence (and thereby knowledge), whether through biotechnology, genetic engineering or AI, any society would do so, or fail to do so at its own peril." Indeed, we see the evidence for the dominant role of intelligence and technology in improving the human condition, in furthering human evolution, leading to what Dick calls the "post-biological universe."

But is that enough? We might consider that cultural evolution can, and perhaps should, lead us to a kind of "post-intelligent," "post-technological" universe—a universe that is not predominantly ruled just by the forces of intelligence and technology, but also by the forces of morality and creativity. Why not? We see evidence for the forces of morality and creativity all around us—and arguably both are increasing at rapid rates.

7.9.1 Caring Capacity and Morality

As a result of our interests as valuing agents, we have emerged in the universe as agents with meaning, purpose, and morality as cultural derivatives of value. If the universe did not have morality prior, it does now. We, in some non-trivial sense, make the universe a moral entity—however limited the degree of that contribution may be. We may indeed be a very small part of the universe that arose by chance, but nevertheless, strictly speaking, the universe now contains morality. The cosmos now has agents caring about other agents and about non-agents as well, and in some cases, about the whole of the universe. This caring, even if it is only a product of biological evolution, need not necessarily be constrained by that narrow biological heritage. Much collective and individual human behavior reflects decreasing violence and an expanding circle of moral consideration for a variety of complex and subtle reasons that arguably go beyond solely mutually beneficial "transactions" (Singer 1981; Sober and Wilson 1998; Gintis et al. 2005; Pinker 2011; Shermer 2015).

We now have an awareness of our capacity to care, and perhaps the sources of such capacity (i.e. biological and cultural evolution). It seems we are aware in a way no other species is. This awareness, our knowledge, can help mind manage biology. And it does appear to be happening. One often hears the refrain that socials ills are inevitable, that they will always be with us. But why should that be? Can we really be so confident that long-lived intelligent species, perhaps ones like ours that exhibit great moral potential, have neither the will nor capacity to eradicate most, if not all social ills? Such certainty appears to be unfounded.

It may be difficult, if not impossible, to ultimately wind our way out of what might be a "selfishness trap" that prevents us from giving up or significantly moderating our selfishness even if we have compelling reasons and opportunities to do so (Lupisella 2001).[22] It may be difficult, if not impossible to render the implications of competition for resources and the second law of thermodynamics negligible, but perhaps it will eventually be possible. As intelligence and technology carries beings to ever-increasing degrees of well-being and comfort, the cost of caring for others can decrease, helping to make it easier to care for others, resulting in more caring acts and an increase our overall "caring capacity" (Lupisella 2013).

As the cost of caring for others is reduced, we may be able to better pursue the well-being of others as a central organizing principle for cultural evolution, including

---

[22]It is conceivable that we will be able to genetically or cognitively moderate internal selfish predispositions on large scales sooner than we think. If so, humanity will be faced with difficult questions regarding whether such an endeavor should be attempted, and if so, how we should do it. Our strongly selfish natures may in fact prevent us from ever seriously moderating or abandoning our selfish motivations—in part because selfishness is important for individual survival, and also because genetically or cognitively moderating our selfishness may have to be an "all-or-nothing" social agreement to avoid undue advantage for those who choose not to. A consideration regarding this kind of constraint is that a sufficiently small and relatively isolated space community (e.g., on Mars) may be able to conduct such an experiment.

perhaps ultimately, the whole of the universe.[23] Indeed, this points to the possibility of a nearer-term cultural goal: reducing the cost of caring as much as possible—which is happening to some extent incidentally as well as intentionally as part of our social organizing and technological evolution. Perhaps reducing the cost of caring, as well as increasing its benefits, can serve as a more explicit formal organizing principle for long-term cultural evolution.

If we are on our own, if there is no larger objective source of value and meaning beyond that which we create, we may then see ourselves simultaneously in opposition to a meaningless and hostile universe (i.e., in opposition to the second law of thermodynamics) and also as a kind of transformative force for creating a meaningful and moral universe. In the same way that religions and other philosophical views have often motivated human beings to care beyond narrow considerations, perhaps cosmological worldviews can as well.

7.9.2 Love as an Emergent Property?

> Life and mind are simply a couple of things that happen to happen. Until they don't. I used to imagine that by studying the universe, by peeling it apart figuratively and literally, we would answer enough of the how questions to catch a glimpse of the answers to the whys. But the more we learn, the more that stance seems to face in the wrong direction. Looking for the universe to hug us, its transient conscious squatters, is understandable, but that's just not what the universe does.
>
> —Brian Greene, *Until the End of Time*

> For small creatures such as we, the vastness is bearable only through love.
>
> —Carl Sagan, *Contact*

If, or when, our "caring capacity" has been reached, if the well-being of all has been sufficiently achieved, what then? Perhaps it is premature (or unnecessary!) to think beyond that, but in some parts of the world, we are increasingly able to achieve substantial well-being of many members of society—although admittedly falling well short of adequate overall global caring capacities. Nevertheless, we see signs of increased caring capacity and perhaps even a kind of deep love emerging in the human species. Indeed, love may become (if it is not already) a critical emergent property of the universe—one that intelligent cultural beings are increasingly bringing forth and realizing to greater and greater degrees. Love may have originated from biological beginnings, but it need not be constrained by its biological origins. Indeed it appears there is potential for love to dramatically transcend biology, and there is reason to think it is already happening.

---

[23]This would be consistent with Gardner's (2009) suggestion that highly advanced intelligence might be guided by a kind of moral cultural attractor that preserves humanity and that might ultimately help the universe as a whole to survive and replicate. This is similar to more recent suggestions by Vidal (2014).

Our increasing awareness of love on many levels can make it more than just an evolutionary product in service of biological fitness. We feel love's force, however diverse and complex it may be. We know its power to motivate, inspire, and create positive human experiences. We know details about the physiological and psychological reality of love, which puts human beings in a unique position to leverage it in the most profound sense. We can take this product of biological evolution to whatever level we seek. We can use love in whatever way we choose—including in a cosmic context. We can choose, however strange or counterintuitive or difficult it may seem to some, to love the whole of the universe, its evolution, and all that implies.

As noted previously, Brian Swimme suggests that the universe is attempting to be felt. Perhaps it is not so much that the universe is actually attempting it, or intentionally trying, but more that is just simply happening unintentionally through a bootstrapped kind of process discussed previously. We do indeed appear to be a way for the universe to be felt—at least in a general sense as feelings emerge in the universe through being like us and as we experience, or directly apply, those feelings in a cosmic context.

In the sense that sufficiently aware beings think of themselves as "feelers" of or for the universe, then in that sense we are intentionally attempting to help the universe be felt. Through our knowledge of cosmic evolution, through specific cultural movements (e.g., such as Epic of Evolution and Universe Story), and through a recognition of our potential roles as cosmic creators and arbiters of value and meaning and purpose for the universe, we are then indeed intentionally helping the universe to not only be felt, but to also have meaning and purpose. We may go further and say that as cosmic reverence emerges in the minds of cultural beings like us, we are way for the universe to be loved if that is what we choose to do. If we choose to love the universe, then the universe is, strictly speaking, loved. We humans then, along with other beings capable of love, are way for the universe to love itself.

Indeed, we can think of love as a real dynamic in cosmic evolution, a kind of cosmic "force." Love is physically embodied in our brains, in the behavior and dynamics of our minds and bodies, in our choices and behaviors. If love reflects what we deeply value, then we will be motivated to act on behalf of what we love. We will care for what we love. We will preserve what we love. We will create what we love. We can choose to cultivate love above all else. We can choose to love our universe. We can choose to love the multiverse if such a thing exists. We can even love future universes yet to be born.

Love is a choice that cultural beings can make. It may be a difficult choice in some respects, an "impractical" and even fanciful choice when applied to something as big and unknown as the universe, let alone multiverse or future universes, but it is a choice nonetheless—a potentially unusually compelling and powerful choice. As we are increasingly able to love and care for all, cultural evolution may point us, take us—we may take it—in directions we ultimately deeply wish for, in directions we ultimately desire and truly love.

7.9.3 Creativity

Ever-increasing degrees of creativity may be one of the directions our cultural evolution takes us. Creativity is a significant, if not a critical part of human culture. The universe itself seems to be highly creative. Even if creativity is not a deeply cosmic phenomenon, it is nevertheless something that human culture (and potentially other cultures) can bring to the universe. Perhaps creativity for the sake of increasing diversity in the universe is a pursuit worthy in and of itself—being mindful that the often advocated idea of "creative destruction" may need to be more carefully considered in light of how much net well-being and diversity is gained, as opposed to the blind justification that sometimes results from its invocation.

It may seem fanciful or gratuitous to think so long-term and so speculatively about the future. But if we consider long-term questions now, it can inform nearer-term pursuits. If our thoughts about long-term norms reveal desirable directions to head in, why not start now? To a significant extent, human beings already do that of course—mostly on shorter timescales. But if we would like to see certain states of affairs in the long-term, presumably we value those things now, today, in this moment. Indeed, many human beings deeply value morality and creativity, which are often magnificently manifested in human culture. These behaviors, in most cases, are presumably not motivated by long-term cosmic perspectives. But perhaps long-term cosmic perspectives can increasingly motivate morally creative pursuits, in part by providing other compelling contexts for those who seek them, including extraterrestrial beings. If they are out there, pursuing a morally creative cosmos may be something we have in common with other beings. If it is not, maybe it should be?[24]

If cultural evolution becomes a major force of cosmic evolution, if the universe undergoes cosmocultural evolution and becomes a deeply "cultural cosmos," then there will be compelling implications for value-driven cultural beings. There will be profound choices to make about our values and their implications. Perhaps we will want to move beyond biology, beyond intelligence, beyond technology, to a universe that is a moral universe, a creative universe—a morally creative cosmos where what matters is not whether cultural beings live effectively, but whether we live morally and creatively with inspiration and love.

Regardless of any deeper truth regarding cultural beings and their cosmic significance, we must still carefully exercise the power of culture. We do not know where it is all heading, and it may be too complex to ever truly understand how it all works,

[24] It is often too easily assumed that extraterrestrial intelligence will have a level of moral advancement that accompanies their technological advancement. While the concept of "caring capacity" covered previously in this book suggests mechanisms for gradually increasing altruism as technology and other cultural constructs evolve, a quick judgment that assumes technically advanced civilizations will be benign to humanity may be unwarranted given the many uncertainties in cultural and moral evolution as well as the uncertainties in perceptions for how different intelligent species perceive each other. Indeed, we may not have to look further than our own species to call this broad assumption into question. See Michaud (2007) and Vakoch (2011) for a consideration of this and related issues.

but culture is carrying us—we are carrying ourselves—on what may be a leading edge of cosmic evolution. It is a wave we have some hand in creating, and we may be more in control than we think. It is an evolution we are partly, if not entirely, responsible for, and we can choose to evolve wisely for ourselves and our universe. Let us then play a worthy role in cosmocultural evolution—a role worthy of our potential, a role worthy of our cosmos.

7.10 Summary of Chapter 7: Cosmocultural Evolution

One way to think about the relationships of cosmos and culture is to explore whether each is important for the other, and if so, how. As noted in the previous chapter on cosmological reverence, unidirectional relationships suggest that the universe is important for culture, but not the reverse. This could be consistent with many worldviews such as a bioresistant, biotolerant, and biofriendly universe, as well as a "weak bootstrapped universe" worldview, which suggests the universe has bootstrapped itself into the realm of value, but without any particular significance for the universe at large.

Bidirectional relationships suggest that the universe is important for culture and that culture is important for the universe. This could include worldviews that can be characterized as a "strong bootstrapped universe," teleological, pantheistic, and theistic—all of which could be consistent with cosmocultural evolution or the cosmocultural principle which suggests that cultural evolution is significant enough for the cosmos that it implies a kind of coevolution of cosmos and culture that should be considered in totality and holistically as single integrated evolution. The new "property" of value that has emerged in the minds of beings with interests and associated intentions, along with the phenomenon of culture that broadly operationalizes value for such beings, has added a significant and arguably qualitatively different kind of evolution to the cosmic landscape.

Bootstrapped cosmocultural evolution suggests that the universe has "bootstrapped" itself into the realm of value via physical processes that created replicators leading eventually to intelligence, mind, and culture which are not necessarily inherent in the universe. Nevertheless, the emergence of mind and culture in cosmic evolution can have limited ontological significance in the sense that such beings can be a means by which an additional property of "value" has emerged in the universe. But bootstrapped cosmocultural evolution can also inform a number of practical cultural considerations such as the importance of biosphereic stewardship, space ecology, off-Earth migration, and SETI considerations.

In stronger versions, bootstrapped cosmocultural evolution may have unlimited potential to eventually influence the whole of the universe itself—including via the five levels of influence that were touched on. The emergence of a new kind of cosmic property, *value*, along with cultural evolution that instantiates value and creates derivatives such as meaning and purpose, along with other endless forms of value, has given rise to a qualitatively different kind of cosmic phenomenon that may have highly open-ended implications and unlimited potential.

What we do with the potentially unlimited power of cultural evolution can be a profound choice, a profound challenge—one that we face day to day on many levels, but that will increasingly be relevant on ever-widening scales as we begin to see ourselves in a long-term cosmic context and as cultural evolution begins to become a more cosmically relevant phenomenon. The forces of morality and creativity can give rise to a *morally creative cosmos*, a universe that goes beyond intelligence and technology, a universe that is deeply driven by the caring capacity of valuing agents and ultimately by a pervasive cosmic force of moral creativity—something to which all cultural beings might ultimately aspire.

Whether one thinks life and culture arose by chance or are instead a part of cosmic design, an argument can be made either way for the value of life, intelligence, and culture. Whether we are random and rare, or part of a larger cosmic dance with other beings, there is arguably some form of noteworthy significance and cosmic value we can claim for life, mind, and culture. Regardless of the details of our origins, we can see ourselves as precious and meaningful, worth developing to whatever potential we envision—for ourselves and the universe.

References

Almar, I. (2002). What could COSPAR do to protect the planetary and space environment? *Advances in Space Research, 30*(6), 1577–1581.

Alston, W. (1998). Ontological commitments. In S. Laurence & C. Macdonald (Eds.), *Contemporary readings in the foundations of metaphysics*. Oxford: Blackwell.

Armstrong, D. (1989). *Universals: An opinionated introduction*. Boulder, CO: Westview.

Banathy, B. (2000). *Guided evolution of society: A systems view*. New York: Springer. (originally published by Kluwer Academic/Plenum Publishers).

Barrow, J. D., Tipler, F. J. (1986). *The anthropic cosmological principle*. Oxford: Oxford University Press.

Blackmore, S. (1999). *The meme machine*. Oxford: Oxford University Press.

Blackmore, S. (2009). Dangerous memes: What the Pandorans let loose. In S. J. Dick & M. Lupisella (Eds.), *Cosmos and culture: Cultural evolution in a cosmic context*. Washington, DC: NASA History Series.

Bloom, H. (2000). *Global brain: The evolution of mass mind from the big band to the 21st century.* New York: Wiley.

Bunge, M. (1999). *Dictionary of philosophy*. Amherst: Prometheus Books.

Carr, B. (Ed.) (2007). *Universe or multiverse*? Cambridge: Cambridge University Press.

Chaisson, E. (2005). *Epic of evolution: Seven ages of the cosmos*. New York: Columbia University Press.

Ćirković, M. (2004) Forecast for the next Eon: Applied cosmology and the long-term fate of intelligent beings. In *Foundations of physics* (Vol. 34). Netherlands: Springer.

Cobb, J. B., & Griffin, D. R. (1976). *Process theology: An introductory exposition*. London: Westminster Press.

Cockell, C. (2007). *Space on earth: Saving our world by seeking others*. New York: Macmillan.

Cockell, C. (Ed.). (2015). *The meaning of liberty beyond earth*. New York, NY: Springer.

COSPAR Workshop on Ethical Considerations for Planetary Protection in Space Exploration. (2012). *Committee on space research*. Based on workshop at Princeton University, June 8–10, 2010.
Cousins, E. H. (Ed.). (1971). *Process theology: Basic writings of the key thinkers of a major modern movement*. New York: Newman Press.
Csikszentmihalyi, M. (1996). *Creativity: The psychology of discovery and invention*. New York: Harper Collins.
Davies, P. (2007). *Cosmic jackpot: Why our universe is just right for life*. New York: Houghton Mifflin Company.
Davies, P. (2009). Life, mind, and culture as fundamental properties of the universe. In S. J. Dick & M. Lupisella (Eds.), *Cosmos and culture: Cultural evolution in a cosmic context*. Washington, DC: NASA History Series. http://history.nasa.gov/SP-4802.pdf.
Dawkins, R. (1976). *The Selfish gene*. Oxford: Oxford University Press.
De Duve, C. (1995). *Vital dust: Life as a cosmic imperative*. New York: Basic Books.
Deutsch, D. (1997). *The fabric of reality: The science of parallel universes—and its implications*. New York: Penguin Books.
Dick, S. J. (2000). Cosmotheology: Theological implications of the new universe. In S. J. Dick (Ed.), *Many worlds: The new universe, extraterrestrial life, and the theological implications* (pp. 191–210). Philadelphia: Templeton Foundation Press.
Dick, S. J. (2003). Cultural evolution, the Postbiological Universe and SETI. *International Journal of Astrobiology, 2*(1), 65–74.
Dick, S. J. (Ed.). (2015). *The impact of discovering life beyond earth*. Cambridge: Cambridge University Press.
Dick, S. J. (2018). *Astrobiology, discovery, and societal impact*. Cambridge: Cambridge University Press.
Dyson, F. (1988). *Infinite in all directions*. New York: Harper & Row.
Epperly, B. G. (2011). *Process theology: A guide for the perplexed*. London: T&T Clark International.
Fry, I. (2000). *The emergence of life on earth: A historical and scientific overview*. New Brunswick: Rutgers University Press.
Galantai, Z. (2004). Long future and type IV civilizations. *Periodica Polytechnica Social and Management Sciences, 12*(1), 83–89.
Gardner, J. (2003). *Biocosm—The new scientific theory of evolution: Intelligent life is the architect of the universe*. Makawao, Maui, HI: Inner Ocean Publishing.
Gardner, J. (2009). The intelligent universe. In S. J. Dick & M. Lupisella (Eds.), *Cosmos and culture: Cultural evolution in a cosmic context*. Washington, DC: NASA History Series.
Gintis, H., Bowles, S., Boyd, R., Fehr, E. (Eds.) (2005). *Moral sentiments and material interests: The foundations of cooperation in economic life*. Cambridge: MIT Press.
Green, B. (2020). *Until the end of time: Mind, matter, and our search for meaning in an evolving universe*. New York: Alfred A. Knopf.
Guth, A. (1994). Do the laws of physics allow us to create a new universe? In G. Ekspong (Ed.), *The Oskar Klein memorial lectures* (Vol. 2, pp. 71–95). Singapore: World Scientific Publishing.
Haidt, J. (2012). *The righteous mind*. New York: Random.
Harrison, P. (1999). *The elements of pantheism: Understanding the divinity of nature and the universe*. London: Element Books (Later via self-publishing site of Taramac FL: Llumina Press).
Hart, W. D. (1996). Dualism. In S. Guttenplan (Ed.), *A companion to the philosophy of mind* (pp. 265–267). Oxford: Blackwell.
Haught, J. F. (2000). *God after Darwin: A theology of evolution*. Boulder, CO: Westview Press (2nd ed., 2008).
Hubbard, B. M. (1998). *Conscious evolution: Awakening the power of our social potential*. Novato: New World Library.
Kaku, M. (2005). *Parallel worlds: The science of alternative universes and our future in the cosmos*. New York: Doubleday.

Kauffman, S. (1995). *At home in the universe*. Oxford: Oxford University Press.

Levine, M. (1994). *Pantheism: A non-theistic concept of deity*. London: Routledge.

Lloyd, S. (2006). *Programming the universe: A quantum computer scientist takes on the cosmos*. New York: Random House.

Lupisella, M. (1999). *Ensuring the scientific integrity of possible Martian Life*. Paper IAA-99-IAA.13.1.08 presented at the International Astronautical Federation Congress. American Institute of Aeronautics and Astronautics, Amsterdam.

Lupisella, M. (2001). Participant statement. In *Humanity 3000 Seminar No. 3 Proceedings. Foundation for the Future*. Seattle, Washington, USA (p. 37), August 12–14, 2001.

Lupisella, M. (2009a). Cosmocultural evolution: The coevolution of cosmos and culture and the creation of cosmic value. In S. J. Dick & M. L. Lupisella (Eds.), *Cosmos and culture: Cultural evolution in a cosmic context*. Washington, DC: NASA History Series.

Lupisella, M. (2009b). The search for extraterrestrial life: Epistemology, ethics, & worldviews. In C Bertka (Ed.), *Exploring the origin, extent, and future of life*. Cambridge: Cambridge University Press (Based on American Association for the Advancement of Science workshops).

Lupisella, M. (2011). Pragmatism, cosmocentrism, and proportional consultation for communication with extraterrestrial intelligence. In D. A. Vakoch (Ed.), *Communication with extraterrestrial intelligence*. Albany: State University of New York Press.

Lupisella, M. (2013). Caring capacity and cosmocultural evolution: Potential mechanisms for advanced altruism. In D. A. Vakoch (Ed.), *Extraterrestrial altruism*. Berlin: Springer.

Lupisella, M. (2016). Cosmological theories of value: Relationalism and connectedness as foundations for cosmic creativity. In J. S. J. Schwartz & T. Milligan (Eds.), *The ethics of space exploration*., Space and Policy series Berlin: Springer.

Lupisella, M., Glenn, J., Jones, C., Dator, J., Dewar, J., Fromkin, D., Ryzenko, J., et al. (2003). *The horizons project: Global mechanisms for long-term survival and development*. Paper IAA-13.2.09 presented at International Astronautical Congress, Bremen.

MacIntyre, A. (1967). Pantheism. In P. Edwards (Ed.), *Encyclopedia of philosophy*. New York: Macmillan and Free Press.

Manson, N. A. (Ed.). (2003). *God and design: The teleological argument and modern science*. New York: Routledge.

Mayr, E. (1992). The idea of teleology. *Journal of the History of Ideas, 53*, 117–135.

Michaud, M. A. G. (2007). *Contact with Alien Civilizations: Our hopes and fears about encountering extraterrestrials*. New York: Springer.

Pinker, S. (2011). *The better angels of our nature: Why violence has declined*. New York, NY: Penguin Books.

Rees, M. (1997). *Before the beginning: Our universe and others*. New York: Perseus Books.

Rolston, H. (1990). The preservation of natural value in the solar system. In E. C. Hargrove (Ed.), *Beyond spaceship earth: Environmental ethics and the solar system*. San Francisco: Sierra Club Books.

Sagan, C. (1973). *Cosmic connection: An extraterrestrial perspective*. New York: Doubleday and Company (2nd ed. published as *Carl Sagan's Cosmic Connection*, 2000, Cambridge University Press).

Sagan, C. (1994). *Pale blue dot: A vision of the human future in space*. New York: Ballantine Books.

Schwartz, J. J. C. (2017). Myth-free space advocacy part I—The myth of innate exploratory and migratory urges. *Acta Astronautica, 137*, 450–460.

Shermer, M. (2015). *The moral arc: How science and reason lead humanity toward truth, justice, and freedom*. New York: Henry Holt.

Shostak, S. (2009). The value of L and the cosmic bottleneck. In S. J. Dick & M. Lupisella (Eds.), *Cosmos and culture: cultural evolution in a cosmic context*. Washington, DC: NASA History Series.

Singer, P. (1981). *The expanding circle: Ethics and sociobiology*. Oxford: Oxford University Press.

Smart, J. (2009). Evo Devo universe? A framework for speculations on cosmic culture. In S. J. Dick & M. Lupisella (Eds.), *Cosmos and culture: Cultural evolution in a cosmic context*. Washington, DC: NASA History Series.
Smith, K. (2008). The terrestrial lifeboat project: An international undertaking to safeguard humanity. In Y. An (Ed.), *A new perspective on applied ethics* (pp. 224–238). China: Renmin Press (in Chinese—English translation available at author's researchgate site).
Smith, K. C. (2009). The trouble with intrinsic value: A primer for astrobiology. In C. Bertka (Ed.), *Exploring the origin, extent, and future of life: philosophical, ethical and theological perspectives* (pp. 261–280). Cambridge: Cambridge University Press.
Smolin, L. (1997). *The life of the cosmos*. New York: Oxford University Press.
Sober, E., & Wilson, D. S. (1998). *Unto others: The evolution and psychology of unselfish behavior*. Cambridge: Harvard University Press.
Swimme, B. (1995). *The hidden heart of the cosmos: Humanity and the new story*. Maryknoll, NY: Orbis Books.
Swimme, B., & Berry, T. (1992). *The universe story*. New York: HarperCollins Publishers.
Teilhard De Chardin, P. (1955). *The phenomenon of man*. Originally published as *Le Phenomene Humain*. Editions du Seuil, Paris (Trans. by Bernard Wall. New York: Harper & Row, 1959).
Tipler, F. (1994). *The physics of immortality*. New York: Doubleday.
Tough, A. (1991). *Crucial questions about the future*. Lanham: University Press of America.
Vakoch, D. (2004). The art and science of interstellar message composition. *Leonardo, 37*, 33–34.
Vakoch, D. (2008). Representing culture in interstellar messages. *Acta Astronautica, 63*, 657–664.
Vakoch, D. (2009). Encoding our origins: Communicating the evolutionary epic in interstellar messages. In S. J. Dick & M. Lupisella (Eds.), *Cosmos and culture: Cultural evolution in a cosmic context*. Washington, DC: NASA History Series.
Vakoch, D. A. (Ed.). (2011). *Communication with extraterrestrial intelligence*. Albany: State University of New York Press.
van Inwagen, P. (1998). The nature of metaphysics. In S. Laurence & C. Macdonald (Eds.), *Contemporary readings in the foundations of metaphysics*. Oxford: Blackwell.
Vidal, C. (2014). *The beginning and the end: The meaning of life in a cosmological perspective*. New York: Springer.
Ward, P. D., & Brownlee, D. (2000). *Rare earth: Why complex life is uncommon in the universe*. New York: Copernicus Books.
Webb, S. (2002). *If the universe is teeming with Aliens .... Where is everybody? Fifty solutions to Fermi's Paradox and the problem of extraterrestrial life*. New York: Copernicus Books.
Wheeler, J. (1988). World as a system self-synthesized by quantum networking. *IBM Journal of Research and Development, 32*(1).
Wheeler, J. (1989). Information, physics, quantum: The search for links. In *Proceedings of the 3rd International Symposium on the Foundations of Quantum Mechanics, Tokyo*.
Williamson, M. (2006). *Space: The fragile frontier*. Reston, VA: American Institute of Aeronautics and Astronautics.
Wright, R. (2001). *Nonzero: The logic of human destiny*. New York: Vintage Books.

Chapter 8
The Connection-Action Principle

8.1 Background and Overview

The two views previously explored, *cosmological reverence* and *cosmocultural evolution*, can both be seen as "soft" or "weaker" relational cosmological views in the sense that they do not make particular ontological or strong metaphysical claims regarding the role of relations or the source of value in a cosmic context, beyond that which is in the minds of valuing agents. As shown in Fig. 1.1, the Connection-Action Principle (CAP) and related ideas are on the stronger end of the relationalist framework and hence support stronger, including more "objective", cosmological theories of value.

The *connection-action principle* presented here can be interpreted to imply a form of "cosmological intrinsic value," perhaps based on an open-ended but still generally "purposeful" and perhaps "trend-like" dynamic in the universe that is implied by stronger versions of the principle. The connection-action principle is suggested to be at least consistent with, if not a potential general explanation for many physical phenomena and philosophical ideas in metaphysics and value theory. Indeed it may be that some kind of metaphysics could at least be helpful, if not needed, to shed light on "boundary questions" (which we should acknowlege may not have any answers at all) such as why the universe exists, why there appear to be physical laws, and how value, meaning, and ethics may be bound up in it all (Murphy and Ellis 1996).

8.2 The Connection-Action Principle

8.2.1 Creativity, Laws, Action

A key notion touched on previously, as part of Whitehead's original articulation of process philosophy, is *creativity*. As covered in previous chapters, the science of

M. Lupisella, *Cosmological Theories of Value*, Space and Society,
https://doi.org/10.1007/978-3-030-25339-4_8

cosmic evolution seems to suggest a highly "creative"—perhaps increasingly creative—evolving universe. How does that creativity arise? We can always appeal to a scientific brute-fact kind of explanation, perhaps consistent with a scientific minimalist perspective. That is, the laws of physics and/or initial conditions of the universe are simply fundamental facts to accept without cause or explanation, and those laws and conditions explain why the universe is creative. More specifically, an initial high-energy, low entropy state naturally gave rise to expansion,[1] with cooling and "clumping" emerging over time—consistent with the second law of thermodynamics, gravity and other physical forces—causing the aggregation or "creation" of objects like atoms, stars, galaxies, planets, and life.

But we can still ask, as many have, why those initial conditions? Why these laws? Even more challenging, why any order or "laws" at all? Why an origin at all, and was it truly from "nothing" as some suggest?[2] Merely being able to ask these questions does not mean they are well-posed or that they have answers, let alone scientific answers, but there have been intriguing and compelling suggestions such as: (1) design by a god or gods, or some kind of entity or beings, including the possibility that our universe is a simulation of sorts (Bostrom 2003), (2) anthropic principles (e.g., Barrow and Tipler 1986), (3) an eternally oscillating universe, going back to the Greeks and forms of Eastern worldviews such as Hinduism and now by some in modern cosmology (e.g., Steinhardt and Turok 2002a, b), (4) cosmological natural selection (Smolin 1997), (5) a metaverse or multiverse that suggests the possibility of many universes (Tegmark 2003), and (6) even more provocative versions of anthropocentric thinking that suggest conscious beings in some sense create the universe and possibly even its laws via extreme interpretations of quantum theory—as explored previously (von Neumann 1932; Wheeler 1990; Davies 2009).

Regardless of the kind of explanation for our universe's origin and its particular laws and initial conditions, most suggestions seem to rest on, assume, or imply that our universe is dynamic, that reality is action-laden. Indeed, it appears we live in a universe of *action*—and action is central to our understanding of our world (Turchin 1993).[3] Even contemplations of an origin as a quantum fluctuation from

---

[1]The initial expansion may have included an extremely rapid and unusually accelerated "inflationary" expansion phase suggested by inflationary theory (Guth and Steinhardt 1984; Linde 1994; Guth 1998).

[2]See Krauss (2012) for a recent scientific exploration of an origin from nothing, but which nevertheless seems to fall short of explaining the emergence of the universe from truly nothing—at least in the traditional philosophical sense of "nothing" (itself a premise Krauss appears to challenge).

[3]Turchin (1993) explicitly links the epistemological criticality of action with an action ontology. Mermin (2017, p. 89) emphasizes action when he writes of QBism: "in QBism, on the other hand, a measurement can be *any* action taken by any user on her external world. The outcome of the measurement is the experience the world induces back in that particular user, through its response to her action." This is also arguably similar to relational quantum mechanics in the sense that it emphasizes the tight relationship between observer and the "external world" and generalizes to all observers and so could arguably be seen as a subset of RQM which, as noted prior, generalizes relational dependencies beyond observers, or "agents" or "users," to *all* objects/systems in the world. We might also note here an apparent consistency with Paul Dirac's "*Interaction* Picture" of quantum mechanics that was also on touched on prior.

a quantum vacuum state or quantum "foam" (a realm of virtual particles which are wavelike fluctuations in the quantum vacuum at "absolute zero") seem to imply there is still "something" that is "dynamic"—that there is at least some kind of action. The quantum vacuum state appears to be a dynamic realm of action, or at least a realm of potentiality that gives rise to action. We seem to live in a fundamentally dynamic physical reality—in a universe full of *action*. But then, of course, we can ask: Why is there any action to begin with?

8.2.2 Connectedness, Relations, Action

In wondering why there is action in the universe, we can again appeal to a kind of scientific brute-fact explanation that suggests our particular laws of physics cause, or at least facilitate or permit, forms of action, without any other explanation. But presumably, there can be action without particular laws per se.[4] An alternative—and admittedly highly conceptual and speculative way to think about a source of action—is to appeal to the oft-cited "connectedness" of the universe. The connected unity of the universe is a theme often embraced by many views such as interpretations of quantum mechanics noted previously (including explicitly "holistic" views of quantum theory (e.g., Bohm 1980; Bohm, and Hiley 1993)), environmental movements (Leopold 1966), and cosmic evolution more generally (Sciama 1959).

As is often the case with many terms and concepts, it can be difficult to precisely define "connectedness." Minimally, connectedness would seem to imply some kind of relationship. If there is a connection, regardless of the kind and however abstract it may be, it seems a relationship of some form would be required to make that connection relevant or meaningful, to realize and instantiate it—to ultimately make it, in some sense, "actionable" in the world. We may be able to use connectedness and relationality almost interchangeably since they are so close in this articulation. However, connectedness might be better seen as a somewhat broader, more fundamental, or slightly different construct (e.g., as a fundamental general *property* that can perhaps be both abstract and concrete), for which relationality is one of what may be multiple forms of connectedness. The basic suggestion here is that the property of connectedness is realized, in part, if not completely, through relations.

Further, for a relation to be fully realized or instantiated, it could be argued that something needs to *happen*. An *event or dynamic* is needed to manifest a relationship, to actualize it. Dynamic events realizing and manifesting relations (or relationships) seems consistent with most forms of process philosophy. Indeed, Whitehead's primary metaphysical construct is "actual events." What we seem to presently know about the universe suggests that even the most "static" phenomena are not truly

[4]Indeed, the very early universe may have been highly chaotic, without much, if any, "normal" law-like behavior (Linde 1986). Wheeler (1983) questions the traditional notion of physical laws in general.

static—including the "emptiness" of a putative quantum vacuum as noted previously. Might this be because relationality, as a form or manifestation of the basic property of connectedness, essentially drives toward some form of action in order to realize relationships and ultimately the general fundamental property of connectedness? Otherwise, how would relations be realized without some forms of "action" or "events" to actualize the underlying relations?

Connectedness giving rise to action might be called the "connection-action principle"(CAP). The most basic or minimalist version of the connection-action principle can be summed up as: *Connectedness is manifested as action.* Put into the parlance of philosophical property talk, it might be stated as: *The universe's property of connectedness is instantiated by action.*

The universe's property of connectedness can also be thought of as a sort of latent "relational potential," and actions are a manifestation of that potential[5] In this sense, relationalism is also foundational because connectedness is relational and actions are relational. Relations are more specific than connectedness but less concrete than actions. This may make "relationality" or "relationalism" a useful bridge or framework in which to understand the connection-action principle. For example, a slightly more specific and/or perhaps slightly stronger version of the connection-action principle might be something like: *connectedness requires relations which are realized as actions*—as suggested by the simple graphic in Fig. 8.1.[6]

The basic "minimalist" version of the connection-action principle does not say that the property of connectedness is necessarily manifested *only* as action. It suggests primarily that action is an important, perhaps critical, manifestation of connectedness and relationality, allowing for the possibility of other forms of instantiations and realizations of connectedness. However, the basic version of the connection-action principle explored here does imply that action is a robust manifestation of connectedness.

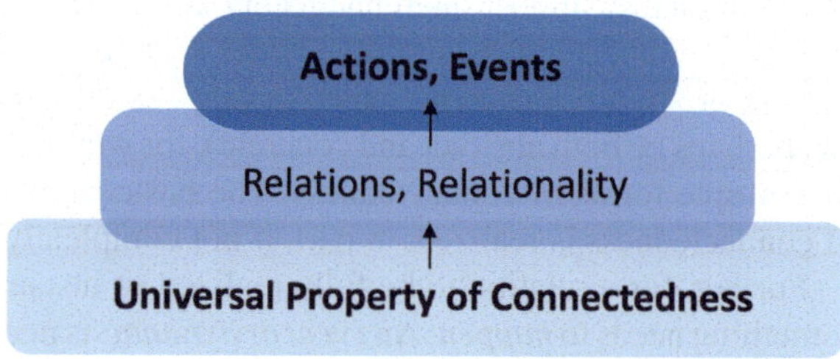

Fig. 8.1 Graphical representation of the connection-action principle

[5]Quante and Engelhard 2018 provide a variety of perspectives regarding "potentiality" and ways to use words such as 'manifest', 'instantiate', 'actualize', and 'realize'—all of which are essentially used interchangeably here.

[6]At the risk of muddying the waters further, a truncated version of the connection-action principle which emphasizes the fundamental role of relations (or perhaps equates connectedness with relations, or relatively de-emphasizes connectedness) could be characterized as a "relation-action principle."

8.2.3 Degree and Necessity: Stronger Versions of the Connection-Action Principle

If we allow for the general possibility that possessing a property can lead to increasing degrees of its instantiation,[7] a stronger version of the connection-action principle can be interpreted to suggest that the universe's property of connectedness could be manifested as *increasing degrees* of action, perhaps *ever*-increasing degrees of action. Increasing degrees of action could be manifested in many ways and is arguably consistent with many forms of process philosophy.

Process can be seen as a form of, or possibly synonymous with, *action*. Increasing realization of connectedness via relations and actions can give rise to more and novel "processes" in the form of increasing creativity, increasing diversity, increasing complexity—continually driving the emergence of new relations, new actions, new dynamics, new phenomena. On this interpretation, increasing degrees of relationality and action (e.g. increasing degrees of diversity and novelty), are a part of the universe increasingly realizings its nature.

The more the universe produces action, the more the universe's property of connectedness is realized—the more its potential is realized. This arguably implies a soft form of teleology, suggesting the universe is, at least in part, about realizing connectedness, and in so doing, it is evolving in many highly dynamic ways, always bringing forth new creations, new relations, and *increasing* novelty and diversity as suggested by certain forms of process philosophy and perhaps even the "temporalized" version of the principle of plenitude (Lovejoy 1936). So a broader, stronger version of the connection-action principle might suggest that the universe's property of connectedness is realized through relations as *increasing* degrees of action. The greater the degrees of relationship and action, the greater the degree to which the universe's "relational potential" is realized.

But why "connectedness"? One start at supporting connectedness as a fundamental property of the universe is to ask how we might conceive of the universe as being fundamentally disconnected. What would that look like? How would such disconnectedness originate? Can we give an account of how even very distant parts of the universe are fundamentally disconnected or unrelated? The Big Bang model (including the observed homogeneity of the early universe) along with physical phenomena such as energy, gravity, and aspects of quantum mechanics/quantum field theory and quantum entanglement (Esfeld 2016) seem to ultimately imply a deep fundamental physical connectedness, relationality, and dynamism in the universe as a whole. For example, gravity and other fundamental forces such as the strong nuclear force literally *connect* physical material together.[8]

[7]This is an important consideration that will not be explored in detail here. Instead, it is suggested that it is reasonable to consider that to possess a property, can, in some cases, result in *increasing degrees* of realizing that property over time.

[8]For example, stars are highly active, highly dynamic entities, essentially created by the "connecting" force of gravity. Stars are literally "turned on," turned into high-energy states, in part by the connecting force of gravity, which then release their radiation and building blocks out into the

It appears it would be extremely difficult, if not impossible, to describe how the universe could be fundamentally internally disconnected—despite what might have been a highly inflationary phase that could have caused different parts of the early universe to move away from each other at extremely high speeds. Such a phase still permits the most distant parts of the universe to be connected. Any attempt to describe disconnectedness would seem to ultimately involve, or be traceable as, a chain of causality giving rise to what is incorrectly seen as fundamentally disconnected parts of the universe that were once at least causally connected. For example, although "baby universes" being "pinched off" from their parent universe might seem truly disconnected, there is still at least a "temporally" connected physical chain of actions giving rise to the newly created universe—which, notably, may have similar properties to its parent universe (Smolin 1997). In any cyclical models of the universe or universes, connectedness would also seem to apply across the various cyclical boundaries.

Even if there have been multiple, or an infinite number of separate universal origins from a more fundamental state of reality such as a quantum vacuum, quantum foam, or quantum field, such universes would apparently still arise from some common base of reality (which might also be called the "metaverse") as opposed to a true "state of nothingness" in which universes come into existence without any connection to any other reality of any kind. Regardless, independently arising universes still would appear to be connected or related in the broad sense of having originated from the same fundamental base of reality, perhaps even with a possibility that there could even be interactions between coexisting universes (Mersini-Houghton and Holman 2009[9]; Mackenzie et al. 2017; Alonso-Serrano et al. 2013) or perhaps between "branes" (Khoury et al. 2001; Webb 2004).[10]

As a general point then, it seems very difficult, if not impossible, to provide a compelling account of fundamental disconnectedness in our universe, or any universe, or any collection of universes.[11] It may be that it is logically necessary for the whole of reality to be completely connected—even if it is connected in a somewhat vague manner. It may be that it is physically, or even conceptually impossible for universes, or any kind of fundamental reality in general, to be anything other than what is ultimately a fundamentally deeply connected unity of some kind—however tenuous or unusual that connectedness may be.

Given what appears to be an empirical and perhaps conceptual or logical basis for deep connectivity of our universe, and given that it may not be possible to provide a

---

universe via highly dynamic explosions (or supernovae) which are part of further creative dynamics such as providing energy, causing mutations, and becoming the molecules of our bodies. Indeed, it seems we are, in many critical ways, the direct result of star dynamics, of "star stuff", as Carl Sagan and many others have poignantly observed.

[9]Kashlinksy et al. (2008) provide observational data that some have interpreted as another universe interacting with ours, e.g., "tilting" it.

[10]We can also make note of the "connectedness" of potentially interacting parallel universes under the "many-worlds" interpretation (Hall et al. 2014).

[11]Sciama (1959) devotes a book to demonstrating that the universe is a connected unity.

sufficient account of fundamental disconnectedness, can we suggest that connectedness is a *necessary* property of the universe? If this reasoning holds, then a stronger version of the connection-action principle could also suggest that connectedness is a *necessary* property of the universe. If relations and actions are ultimately required or necessary to realize connectedness, then a stronger, fuller version of CAP might be: *The universe's necessary property of connectedness requires relations which are necessarily manifested as increasing degrees of action.*

So, the greater the degree of action,[12] the greater the degree to which the universe's fundamental necessary nature, or potential, is realized. Whether a basic characteristic or property of the universe could, should, or must be realized to ever-increasing degrees is not necessarily required by weaker versions of the connection-action principle. But stronger versions would suggest that kind of necessity and trend. The logical possibility that the property of connectedness could be realized as a trend toward increasing degrees of actualizing connectedness through actions may be enough for our general purpose here to articulate plausible stronger versions of CAP and the potential trend-like implications. However, there are likely to be more specific ways to infer a trend toward increasing degrees of action from a universal property of connectedness.

One possible approach for CAP to more specifically imply a trend is to consider that increasingly robust forms of relations and actions result from building upon the old to create the new,[13] since otherwise, there would only be mere repetition (an idea consistent with process philosophy). This construction or creativity built upon what has come before can lead to emergence and the emergence of increasingly complex systems, which can then be continually built upon to endlessly create novel and increasingly robust forms of relationships and action.

8.2.4 *Hints of Intra-action and Relational Metaphysics*

The connection-action principle (CAP) is consistent with process philosophy in that it suggests a conceptual or "metaphysical foundation" that gives rise to processes, events, actions. An even stronger version of CAP can be interpreted to make an ontological claim that relations are all that really exists (Oliver 1981). Interpretations of this kind can be seen to support the idea that many conceptions of philosophical substantivism may be a kind of illusion—as suggested by a number of philosophical traditions touched on previously in the chapter on relationalism.

---

[12] What constitutes "greater" action has been briefly touched on in the very general sense of suggesting characteristics and qualities such as emergence, creativity, complexity, self-organization, diversity, etc. But this is a very open-ended question, and "greater" action could presumably take many forms, which could be explored in subsequent work, perhaps even with quantified analyses.

[13] Volk (2017) has recently used the term "combogenesis" to emphasize how simpler smaller entities have continuously been brought together throughout cosmic evolution to create new larger forms with new sets of relations.

However, in its stripped-down "minimalist" form, the connection-action principle can be interpreted to be agnostic regarding whether the relations and interactions are between material objects (materialism/substantivism) or *intra*-actions (Barad 2007) within a highly blended web of relations that may not require any particular material substantive objects at all. Nevertheless, an interpretation that only relations and process are ultimately required for knowledge (e.g., the relational epistemology of relativity and quantum theory covered previously), combined with a consideration that material objects do not appear to be ontologically necessary, suggests it is reasonable to interpret the connection-action principle as incorporating, or at least implying, a form of *intra*-action as opposed to interaction among discrete material entities.[14]

It could also be argued that *intra*-action allows for a more robust realization of the connection-action principle in the sense that it is ultimately relations and actions "all the way down"[15] and "all the way across." There is no discrete, material, substantively "atomic" end point at which relationality and action are ultimately constrained. There is no "point" at which relations and actions stop in a single instance of irreducible actionless matter. A relational metaphysics of this kind (or "relation-action" metaphysics more specifically) arguably allows for more "depth," or more opportunity for more relationality and action, perhaps of infinite depth. Indeed, the history of particle physics seems to reflect a kind of endless depth to "particles"—including in the sense that quantum field theory sees particles primarily as transient waves in seamless fields. So under stronger interpretations of CAP, *intra*-action arguably implies more potential for realizing the nature of the universe.

Interaction, or intra-action, and a relational metaphysics more broadly, could be further explored in more detail by leveraging fields such as systems theory, cybernetics, complexity theory, modern network theory, and action ontology (Turchin 1993; Heylighen 2010; Heylighen and Beigi 2018)—all of which have overlapping and interrelated subsets that could allow for more detailed exploration (and perhaps some degree of formalization) of the kind of metaphysics implied by the connection-action principle. Arguably, the connection-action principle provides a conceptual foundation for why there should be an ontology of actions to begin with. More detailed applications leveraging the above areas are beyond the scope of this book but could be examined more thoroughly in follow-on treatments. Related to that suggestion, what follows is an attempt to at least consider briefly how information and complexity relate to the connection-action principle.

---

[14]This is also arguably consistent with interpretations of quantum mechanics such as David Mermin's suggestion that only correlations are ultimately real, that reality is constituted by "correlations without correlate" (Mermin 1998, 765).

[15]For a contemporary refutation, see Briceño and Mumford (2016)

8.3 Emergence, Biology, Information, Complexity

Theories of emergence and complexity may ultimately have much to say about how increasing degrees of action can be more precisely measured and how complexity might be linked to conceptual models such as the connection-action principle. A trend of increasing complexity can at least conceptually follow from increasing degrees of action because complexity, with its diverse and large number of connections and relationships, as well as high information content, can be said to correspond to a high degree of *inter/intra* action.

The importance of connections and inter/intra action is also realized by considering the following view regarding the origin of life. Stuart Kauffman writes:

> As the diversity of molecules in our system increases, the ratio of reaction to chemicals, or edges to nodes becomes ever higher. In other words, the reaction graph has ever more lines *connecting* (italics added) the chemical dots. The molecules in the system are themselves candidates to be able to catalyze the reactions by which the molecules themselves are formed. As the ratio of reaction to chemicals increases, the number of reactions that are catalyzed by the molecules in the system increases. When the number of catalyzed reactions is about equal to the number of chemical dots, a giant catalyzed reaction web forms, and a collectively autocatalytic system snaps into existence. A living metabolism crystallizes. Life emerges as a phase transition. (Kauffman 1995, 62)

The conceptual key to this account is that the ratio of *connections* (in this case, interactions in the form of chemical reactions) to other elements in the system (in this case molecules) is what causes a phase change to "emerge." In an important sense, the system's degree of connectedness is fundamental and ultimately gives rise to the emergence of new dynamics.

The connection-action principle is consistent with this kind of account in the sense that it implies that what is important are connections in a system which ultimately drive internal relations and actions of the system. It might even be that we can see this connectedness and relationality of a prebiotic system as a beginning for how "software" might emerge from "hardware"—something Paul Davies suggests is an important transition for the emergence of life (Davies 1999). Connections and the associated relations and interactions within a system arguably represent some relative informational state within the system,[16] so a high degree of connectedness within a system can also be seen as representing a high degree of information content.

On this informational view, there may not be a fundamental conceptual problem of having to make a transition from hardware to software since the universe is in some important sense already primarily software, or information, or perhaps computation (Wolfram 2002; Lloyd 2006; Davies and Gregersen 2010; Bokulich and Jaeger 2010) in the general forms of connections, relations, and interaction or intra-action. There are no concrete "things" necessarily, just information, or more generally, relations.[17] In this context, life would be a matter of degree of actualization of the universe's

[16]This is similar to the relational quantum mechanics interpretation we explored previously.

[17]For a balanced and cautious view on using information is a fundamental construct, see Timpson (2010).

informational/relational nature, as opposed to an unlikely or radical emergent phase transition that appears to be dramatically different from the rest of the reality from which it arose. Nevertheless, something like phase transitions (which again, may ultimately be matters of degree, albeit importantly, large jumps in degrees)[18] would still be a way for highly novel emergent phenomena to occur, helping to realize yet more novel action—a way for the universe to explore a very large, if not infinite, possibility space of its potential driven by connectedness realized as action.

As noted previously, complexity is another way to think about the kinds of biological and emergent dynamics touched on above. Numerous connections and any associated actions can be seen as being similar to complexity. Complexity can be a problematic subject (Rescher 1998), but a bottom line here is that in a universe that is realizing connectedness in the form of actions and increasing degrees thereof, complexity is arguably "predicted" in the sense that complexity may be a way for the universe to increasingly manifest more novel forms of relations and actions.[19]

8.4 Holism and the Unending Process of Movement

Keeping Things Whole

In a field
I am the absence
of field.
This is
always the case.
Wherever I am
I am what is missing.

When I walk
I part the air
and always
the air moves in
to fill the spaces
where my body's been.

We all have reasons
for moving.
I move
to keep things whole.

[18]Turchin (1977) explored the idea of a "metasystem transition" to help explain significant transitions.

[19]Rescher (1996, 2006), a proponent of process philosophy, sees nature as continuously evolving process toward increasing complexity.

—Mark Strand [20]

The suggestions in the previous section point to the possibility of thinking about the universe as a kind of ecosystem, perhaps a kind of information ecosystem. Along these lines, the connection-action principle seems to also suggest why the universe is best understood as an integrated collection of relations and actions—as reflected in some relational views explored previously[21] and holistic interpretations of quantum mechanics. David Bohm writes:

> I would say that in my scientific and philosophical work, my main concern has been with understanding the nature of reality in general and of consciousness in particular as a coherent whole, which is never static or complete, but which is in an *unending process of movement and unfoldment* (italics added). Thus, when I look back, I see that even as a child I was fascinated by the puzzle, indeed the mystery, of what is the nature of *movement* (italics added). (Bohm 1980, viii)

Bohm's reflection seems consistent with the motivation of asking why there is action—as well as with the implications of the connection-action principle. Movement and action are arguably very close in meaning, so Bohm's reference to the universe as an *unending process* of *movement* and *unfoldment* is very similar to the implication of the connection-action principle that it gives rise not only to action but, in stronger interpretations, to ever-increasing or unending degrees of action (or movement) over time.

8.5 Laws, Time, Plenitude, Multiverse?

In the context of the connection-action principle, we might see laws (or something like "laws that can act as a fairly consistent and stable framework with which to structure actions), as one way for connectedness and action to be robustly manifested—since some kind of physical consistency, can allow for the relatively stable "construction" of new, perhaps increasingly complex, creations. Laws can permit the new to be built upon the old, and hence new and more robust forms of action can be realized (similar to precepts of many articulations of process philosophy). Otherwise, pure random chaotic action would persist and presumably not allow for the construction or creation of new, increasingly diverse and complex forms. This is arguably a similar conceptual foundation of the evolutionary developmental model touched on briefly in the chapter on precedent—albeit broader and less constrained. That is, some form of developmental constraints (e.g., "laws") are needed to complement and

[20]"Keeping Things Whole" is a poem by Mark Strand published in 1964 that was originally titled "A Reason For Moving" in *Sleeping With One Eye Open*, published by Stone Wall Press (Iowa City). The poem was subsequently included with the title "Keeping Things Whole" in a 1968 publication, *Reasons For Moving*, and also in a 2014 volume, *Collected Poems*, published by Alfred E. Knopf in the year of Strand's death at age 80.

[21]Leibniz and others such as Mach, had a relational view of space. Carol Cleland (1984) also articulates a relational view of space that does not rely on the supervenience of lower-level properties.

create relatively stable complexity, leveraging and building upon the less constrained dynamics of more random and open-ended evolutionary processes (Smart 2009).

If connectedness is realized over time as ever-increasing degrees of action and hence new relations (e.g., increasing diversity and increasing complexity, etc.), does this provide support for the idea of the temporalized version of the principle of plenitude that suggests diversity increases, perhaps to infinity, over time? [22] Could "time" be a way for the universe to increasingly realize connectedness through ongoing, perhaps eternal creativity and emergence? Even if time's manifestation, or its "arrow," comes from the second law of thermodynamics and an initial low entropy state, time can still be seen as a way to give the universe many, maybe infinite, opportunities to realize connectedness through action since there is some form of "duration" that allows continous creativity.

If time's arrow can ultimately be traced back to the Big Bang when there was a low entropy state, we can wonder why the universe might have had that initial low entropy condition (Caroll 2010a, b). One possibility is that our universe is not a closed system and that the initial low entropy state came from a previous universal cycle and/or a manifestation of a larger multiverse or "metaverse" from which universes of initial low entropy emerge either as statistical brute facts or perhaps for some other reason. For example, a cyclical universe and/or multiverse could be a robust manifestation of the most generalized version of the connection-action principle—a strong realization or "instantiation" of a generalized property of connectedness across all of reality, giving rise to more and more action across of all of reality, across all cosmological cycles, across all universes.

So, a multiverse may be a way to realize very high levels of action, particularly when taken in the aggregate. Presumably more universes give rise to more action, which contributes to more creativity, diversity, complexity, etc. The postulated "ground" of reality from which many, possibly infinite, universes may emerge, (e.g., a quantum vacuum state or quantum foam), while still a realm of dynamism and action, can also be interpreted to suggest it is nevertheless a low-action state.[23] Such a minimal state may be seen as a lower or weaker manifestation of connectedness in the sense that it is a relatively homogenous and minimally dynamic state—with, nevertheless, high potential. A homogenous state could be considered a kind of "low connection-action state" since there is low diversity of action and hence less action overall. These lower states of homogenous action may give rise to "higher" states of

---

[22]Deutsch (2011) argues that much of what we think of as fiction is likely to be fact somewhere in the multiverse.

[23]Krauss (2012) suggests that the unstable nature of the quantum vacuum state postulated by quantum field theory is effectively a state of nothingness from which the universe arose via the activities of virtual particles that emerge from that quantum vacuum state. However, there are disagreements as to whether a quantum vacuum state is truly "nothing" (Albert 2012) since there appears to be activity associated with such a state. However, it is not clear that this distinction is significant for our purposes here, with the possible exception that a true state of nothingness might be "unstable" and/or logically impossible if connectedness needs to be manifested by action (or even in some other way).

action with more differentiation and diversity (e.g., the emergence particles or universes) and arguably then more robust realizations of connectedness as more robust action, diversity, and creations emerge.

The connection-action principle may be a conceptual/metaphysical explanation for why the base of reality "moves off" its low base state. Something like a quantum vacuum state, a low energy quantum field, could be the lowest state of action possible, from which higher levels of action spring into existence, perhaps necessarily, "automatically," so that connectedness is better realized as increasingly robust action relative to that base state. Perhaps we can interpret the instability of a quantum vacuum state as a manifestation of the need to push out of the lowest action state into some new, or more robust action state. If the lowest energy quantum field state (e.g., a "quantum vacuum") is the "floor" of reality, maybe it's only "up" from there.

8.6 Value Theory and the Connection-Action Principle

8.6.1 Intrinsic Value Revisited

G. E. Moore says this about intrinsic value: "To say that a kind of value is "intrinsic" means merely that the question whether a thing possesses it, and in what degree it possess it, depends solely on the intrinsic nature of the thing in question." (Moore 1903, 286)

Or, as Lemos puts it: "The intrinsic value of a state of affairs is absolute, in the sense that it is determined completely by the nature of the state of affairs in question taken completely in abstraction from any consideration of the value of other states of affairs that are not constituents, elements, or parts of it." (Lemos 1995, 34)

There are two very important ideas expressed in these conceptions of intrinsic value: independence and "nature-dependence." Moore uses the word "solely" to capture the idea that intrinsic value is internal to thing in that its value is not dependent on anything else but its own nature. Intrinsic value is independent of anything else. Lemos uses the word "absolute" to capture the same idea. Both Lemos and Moore emphasize the importance of grounding value in the nature of a thing, so that it is in virtue of how a thing is, or what a thing is, that gives it its intrinsic value.

In describing intrinsic value, it seems we would want to give an account of the nature of some general state of affairs and why such a state of affairs is intrinsically valuable. So we need both, a description attempting to capture some nature, and more importantly, some account as to why that nature should be intrinsically valuable.

Previously, we briefly noted the possibility that the connection-action principle can be interpreted to imply a kind of intrinsic value, a kind of value intrinsic to the universe, to its nature, to its unfolding through relations and action. If, as the connection-action principle suggests, a universal property of connectedness is realizable through relations and action (and perhaps ever-increasing degrees thereof in stronger versions), then we might think of this as a kind of cosmic grounding of value, or "cosmological intrinsic value," in the sense that the property of connectedness and its manifestation and implications are intrinsic to the nature of the universe and its evolution.

This suggests that the universe may "be about something"—albeit admittedly broad, highly open-ended, and largely non-prescriptive. Nevertheless, if connectedness is a fundamental property of the universe, and if connectedness gives rise to relationship(s), action, creativity, diversity, complexity, etc., then the universe in some very real sense is realizing its nature via these characteristics and dynamics. Can the universe realizing its nature be a sufficient grounding for intrinsic value? If the universe (or multiverse/metaverse) is all that is, all that was, and all that ever will be, then perhaps it can serve as a compelling "objective" framework for grounding and assessing intrinsic value.

If the universe is realizing its nature, then that realization is intrinsic to the universe and hence can be argued as a basis for value that goes well beyond the value created by valuing agents—it is arguably of value to the universe for it to realize its nature or at least part of its nature. But even if this bridging from a "fact" to a value is insufficient to truly ground intrinsic value, we can at least argue that it is a reasonable philosophical *choice* to use the nature of the universe and any associated trend to ground and justify intrinsic value. And even if truly intrinsic value in its purest, strictest sense cannot be completely justified with this account, it is at least arguably a very strong form of value. It is value associated with something significant and compelling: our universe—and so it is value that arguably goes beyond the parochial interests of any particular intelligent agents.

The connection-action principle suggests that relationality in general, and specific relationships and associated actions in particular, are valuable. This was suggested previously in the weaker forms of relationalism, but the connection-action principle makes this kind of "relational value theory" much stronger—it suggests a conceptual basis, if not a metaphysical foundation for instrinsic value in the universe. The connection-action principle suggests what is arguably an objective ontological or cosmological grounding for the value of relations and resulting actions that emerge—a grounding that does not depend on valuing agents, but that is "recognized" by valuing agents (and perhaps in many other ways as well). On this view, all forms of relations and intra/interactions are in some sense, valuable. This is obviously problematic on a number of levels that will be explored briefly. One way to address some challenges of this implication is to explore *degrees* of value or degrees of intrinsic value. The connection-action principle can be interpreted to provide at least a theoretical framework for considering degrees of value.

8.6.2 Degrees of Value

Regardless of the ontological commitments as to whether there are ultimately irreducible atomic "things" vs. just relations, what is valuable in the context of the connection-action principle is realizing connectedness through relations and action in a myriad of forms that may take—many or most of which are presumably yet to be understood or atualized. What is valuable is an unending unfolding of creativity, perhaps leading to ever-increasing degrees of creativity. Can we say then that ever-increasing degrees of connectedness, relations, and action, represent forms or expressions of *degrees* intrinsic value, or degrees of "cosmological intrinsic value"?

Most attempts to deal with the issue of value measurement have focused on value as it pertains to human ends (Handy 1970; Zenzen and Hammer 1978). The philosopher, Alan Marshall, in exploring how extraterrestrial life might be viewed, stresses that intrinsic value is not imposed by humans but is merely recognized by humans (Marshall 1993). This polar tension is indicative of thousands of years of how value has been viewed in such fundamentally different ways and reflects the difference between instrumental value and intrinsic value—an important distinction for which we may see some partial integration going forward.

Treating connectedness, relationality, and action as intrinsically valuable provides what could arguably be a broader objective frame of reference against which to assess value. In theory, the connection-action principle could provide a potentially "scientific," or at least reasonably objective, way to assess value. The degree to which connectedness is realized via action could be seen to correlate with degrees of value or intrinsic value—or at least relative degrees of value. A basic utilitarian intent could still remain in the sense that a kind of "measurement" of what is valuable can be attempted, but the measurement could be informed by something other than solely anthropocentric or biocentric motivations.

Assessing and measuring the value of action would presumably be very complicated and perhaps ultimately untenable for a variety of reasons. What precisely is meant by relationality, action, interaction, or intra-action so that it could be accurately, precisely, and fairly assessed? As was asked previously, what constitutes lesser or greater degrees of relationality and action?

Organized complexity, as a possible way to recognize robust forms of action, is consistent with the idea of a self-organizing universe trending toward ever-increasing degrees of organized complexity. Scientists are attempting to measure complexity and some contemporary philosophers have been reevaluating it to inform ethical thinking (e.g., Smith 2014; Delahaye and Vidal 2018). Organized complexity is arguably a robust actualization of connectedness in the sense that complexity usually involves a large number of diverse relationships and interactions with high information content useful for driving novel and complex interactions, creations, and emergent phenomena. But the assessment of value as action would involve much more than just complexity since it could presumably be manifested in many other ways.

Indeed, connectedness and action are meant here in the very broadest sense. We are not necessarily limiting the notion of action to the kind of physical action we experience every day. It does not mean the universe and everything in it must always be buzzing in a sea of physical action as the only way to manifest connectedness and relationships and hence value. Connectedness, relationality, and action can presumably be realized in many different, unusual, and unknown ways. In effect, the connection-action principle arguably implies the potential realization of a very large, if not infinite possibility space of value.

8.6.3 Diversity and Freedom

The relationship between the connection-action principle and freedom and diversity is something like the relationship between teleological (utilitarian) and deontological (virtue or duty-oriented) ethical theories. Deontological theories can be thought of as shorthand rules of thumb for realizing the greatest good (Altman 1984). Similarly, diversity and freedom can be seen as as rough "rules of thumb" for generating relationships and action since they give rise to the production of novel action in the world. Freedom, with its reduced restrictions on human agency (and other life-forms as well), can be seen as an effective "mechanism" for moving efficiently through a very large possibility space of action to find increasingly rich and complex relationships and interactions. Unique or rare creations might be worth preserving because of how they contribute to diversity.

The connection-action principle, similar to the "principle of the sanctity of existence," also implies that existence is intrinsically valuable (MacNiven 1993) in the sense that the more the universe produces, the more that exists, the more the universe's potential is realized. In its barest form, we again see the consistency with the principle of plenitude, versions of which have been proposed since Plato's Demiurge, a "God" who wanted the world to lack nothing. Giordano Bruno, Spinoza, and Leibniz (in the form of the principle of sufficient reason) have also proposed ideas similar to the principle of plenitude and are the Western parents of Lovejoy's more modern articulation. Nozick (1981) also explores the idea of "fecundity" in way that is similar to the principle of plenitude. The connection-action principle arguably provides a conceptual foundation for why the universe produces diversity.

This kind of account is consistent with those who cite diversity as a goal worthy in and of itself such as Freeman Dyson who writes:

> Diversity is the great gift which life has brought to our planet and may one day bring to the rest of the universe. The preservation and fostering of diversity is the great goal which I would like to see embodied in our ethical principles and in our political actions. (Dyson 1988)

The connection-action principle differs from Dyson's first sentence by suggesting that the universe may bring diversity to itself, with the biosphere of Earth being an important example, perhaps one of many that may exist throughout the universe.[24] Indeed, in making choices consistent with this view, humanity may help encourage diversity here on Earth and throughout the universe.

This kind of view might also assign a significant degree of value to unique nonliving objects such as the Grand Canyon or Valles Marinaris on Mars. We recognize these valuable actualizations to have "formed integrity"—to use Holmes Rolston's phrase (Rolston 1986, 2012)—perhaps in part because they contribute significantly to the diversity of the universe. If the continued existence of such forms does not prevent degrees of action from being realized in other ways, their preservation is presumably justifiable.

---

[24]The large number of exoplanets that have been discovered points to an increasing probability of "Earth-like" planets beyond our solar system.

Freedom provides many paths of action. However, freedom is also separate from diversity in an important sense. Freedom can be seen as a natural uninhibited realization of action—an unfettered path to action. Freedom is in some sense a path of least resistance to creating relationships and actions and as such contributes as a kind of efficient diversity generator. Biological and intelligent agents appear to exhibit and strongly pursue freedom of action, so we might say that life and intelligence can be robust sources of cosmic value—but not at the exclusion of all else. This kind of perspective transcends human and biological biases and extends value to the broadest sphere, while also acknowledging high degrees of value for living things that benefit from freedom of action and as a result may bring additional diversity to the universe.

In general, Thinking about the universe and value this way would arguably encourage us to be careful about how we interact with the rest of the universe. On a common sense consequentialist approach, it would encourage us to exercise thoughtful caution about possible consequences of our actions before we act. It would require us to be responsible and respectful to more than just the human species—again, something many humans are increasingly recognizing intuitively or by exposure to ideas regarding the value of non-human life (Singer 1981; Rollin 1981).

8.6.4 Aesthetic Value and Organic Unity

It is interesting to consider the extent to which a metaethical theory of value could accommodate aesthetic value theory. In the spirit of seeking underlying unifying principles, it can help to have a comprehensive account of value so that it might tie together ethical value theory and aesthetics. We can argue that to some extent, ethical value has a strong, if not critical "aesthetic" component—at least in the sense that perceived ethical value might arise from internal emotional guidance that pulls us toward things we find "appealing" for any variety of reasons (including evolutionary reasons). Pursuing this kind of approach is potentially a way to find common ground between emotive and realist views of values and ethics if those aesthetic principles can be shown to "supervene" on or emerge from mental states which are generally shared by all humans or all beings capable of the relevant experiences.

Robert Nozick draws on the concept of organic unity from aesthetics and uses it to help ground intrinsic value. He writes: "The more diverse the material that gets unified (to a certain degree), the greater the value." (Nozick 1981, 416). This unity in diversity is what Nozick suggests can be equated with intrinsic value. Importantly, he suggests that organic unity might be the best approximation to value because our experience may be limited regarding what is valuable. However, Marshall (2002) provides a provocative deconstruction of both "unity of nature" narratives as well as distinctions between "mechanicism" and "organicism" more generally—arguably calling into question the whole notion of "organic unity" as a useful aesthetic concept, let alone a basis for objective value. Marshall seeks:

"...an alternative postmodern conceptualization of ecology whereby nature itself may be deconstructed in an ecopolitically benign way through an awareness of 'Otherness'. Within such a conceptualization, it is held that we should not be scared of identifying and locating the differences of the various members of the environment; splitting them up and atomizing them, since if we do it with a view to celebrating Otherness then we may well contribute to an environmental narrative that is better equipped to value the individual lives of each living member of the world." (Marshall 2002, 6)

Notably, Marshall's alternative "otherness" narrative can be interpreted to be somewhat consistent with something like "organic unity" in the sense that diverse individual lives can remain discrete and individual, yet be "brought together" in natural ecological contexts with at least some degree of unity, with at least some degree of important interrelatedness within a broader context of ecological system dynamics as well as philosophical constructs.

Although Nozick gives a helpful account of how organic unity fits with our general perception of value, a more objective justification for why organic unity should be considered valuable might be worth pursuing in the context of the connection-action principle. Nozick is careful to distinguish between demonstrating the value of organic unity from demonstrating why there should be value at all. Without going into all the details of Nozick's criteria, the first question as to why organic unity would be valuable is not addressed in a fundamentally objective approach because his criteria for being valuable are ultimately connected to a valuing agent. Nozick admits this dependence when he acknowledges that all we can do is approximate value because our experience may be incomplete.

Regarding objective value, Nozick writes: "The philosopher's quest for a basis for (and a theory of) objective values, to ground them and understand them, itself involves a value judgment: it is better that there be objective values." (Nozick 1981, 434). Regarding why there should be any value at all, Nozick writes: "What is needed to bring value to our universe is our reflexive choice that there be value, our reflexive imputation of the existence of value ... The fundamental choice open to us is whether or not there will be value" (Nozick 1981, 563).

Andres Altman responds with:

"The basic problem with Nozick's explanation of how value arises is that it presupposes the very phenomenon which it is supposed to explain. To suppose that the fundamental value choice can be reflexive in the way Nozick describes is to presuppose that human choice can create objective value. But that is the mystery that needs explanation. How can human choice bestow on any fact (including facts about organic unity) prescribe authority? How can we, by a simple act of will, bestow upon anything the normative power to determine what our preferences and choices ought to be? Nozick's theory does not answer these questions, and by failing to do so, it fails in explaining how organic unity, or any other factual feature of the world, could have the status of objective value." (Altman 1984, 150)

Altman may underestimate the possibility for reflexive choice to produce something that is, in some important sense, "objective". For example, perhaps in the sense that "emergent" social conventions, relations, and properties can arise and become independent of any one person's views, there can be a certain kind of objectivity, or at least limited or partial objectivity of sorts. Altman himself arguably comes close to this suggestion by indicating that social agreement be required for what is objective. However, Altman rightly brings light to the more fundamental challenge of how true objectivity can be fundamentally linked with value.

Regardless of whether organic unity can be considered "objective" value and/or intrinsically valuable, organic unity does arguably follow from the connection-action principle in the sense that organized complexity is similar to organic unity because organized complexity involves the organization (or "unity") of diverse interactions. Indeed, as Nozick points out, biologists often refer to organisms as having organic unity. It may even be that organic unity is redundant in that organic captures both diversity (the diversity enabled by the organic world) as well as unity, in the form of "organization"—a word clearly related to "organic." The connectedness of the universe corresponds to unity, and action creates diversity. The connection-action principle says that diversity (e.g., novel action) comes from the connectedness (e.g., unity) of the universe as it realizes its nature. The connection-action principle appears to at least be consistent with Nozick's use of organic unity, if not an underlying theoretical foundation for it.

8.7 Challenges and "Predictions"

One might ask why the connection-action principle (CAP) rises to the level of being called a principle—a word normally associated with strong sweeping fundamental claims. One response is that a principle can have varying degrees of significance, and that there are not necessarily well-formulated standards for its use in philosophy or other areas. Nevertheless, the connection-action principle does make a fundamental claim with varying degrees of strength and consequence for various versions and their associated implications—as covered previously (e.g. in the section on "Degree and Necessity"). The fundamental claim proposes what is at least a conceptual or metaphysical foundation for why the universe is a relational dynamic entity. CAP does not stipulate a relational metaphysics as a first principle but instead provides a potential conceptual grounding for why there might be a relational metaphysics at all—namely that the fundamental property of connectedness gives rise to relationality of some form, which is then instantiated as many forms of action—perhaps *necessarily* and perhaps in ever-increasing degrees (in stronger versions touched on previously).

The connection-action principle is a specific fundamental theoretical proposal about the nature of reality that is arguably consistent with, if not a foundational explanation of, the relational nature of fundamental science such as relativity and quantum theory, and perhaps even informational models of reality that are recieving increasing attention. So, CAP is simultaneously a specific form of a relational metaphysics as well as an explanation for why a relational metaphysics might exist at all. CAP also has potential for broad applicability—including applicability to value theory—which was touched on previously and will be explored further in the next chapter.

But of course there are challenges, for example: (a) the incompleteness of the science and philosophy that attempts to defend the idea of connectedness as a fundamental property of the universe, (b) questioning the linkage that relations and actions follow from connectedness, (c) questioning the need to invoke a conceptual abstraction to explain metaphysical "why" questions that may be misguided, unanswerable, or untestable, (and for which much science may suffice anyway), (d) inferring value, and possibly intrinsic value, from what may be tenuous conceptual speculations that could also have problematic consequences and be misused, and (e) a high degree of generality or variability and a lack of specific "measurable" practical utility. Many of these concerns have been at least partially addressed previously, and some will be addressed further in the next chapter, but not all of these challenges will be addressed further here. Instead, we will focus primarily on the potential for the connection-action principle to have some practical operational utility. Additional utility for cosmological theories of value and cosmic ethics will be explored further in the following chapter.

Admittedly, it is not clear how to make an idea like CAP sufficiently scientifically operational and practical enough for some theoretical mathematics or concrete measurable predictions to be credibly developed. But there are ways to think about how CAP might be consistent and possibly explanatory with respect to numerous scientific theories and philosophical ideas—much of which has already been suggested. One significant "epistemological" challenge with CAP is that it can be interpreted to be consistent with almost any observations and predictions involving relationships, action, interaction and/or intra-action, creativity, diversity, complexity, and increasing degrees thereof—which is arguably applicable to "everything" and too general to be specifically practical and useful. Indeed, one way to think about a good explanation is that it cannot be varied too much. [25] Also, as touched on previously, how would we "measure" implications of CAP in a consistent and credible way that could inform questions of value? The practical precision that might ultimately be desirable for CAP is something that can be further explored in future work, but we might consider some general predictions that CAP may imply, some of which are consistent with some views explored in Chap. 2 and throughout the rest of this book.

8.7.1 "Predictions"

At the risk of wading even deeper into what are already speculative waters, let us briefly consider some general possibilities and perhaps what could be considered high level and admittedly somewhat vague "predictions" that might follow from the connection-action principle:

1. Increasingly robust degrees of interaction and/or intra-action will be discovered and will emerge over time—e.g., life and intelligence may be found throughout the universe and/or dynamic forms similar to life may emerge over time.

[25] Deutsch (2011) cautions against theories that are too easy to vary.

2. Additional forms of "deep connectivity" will be discovered as further instantiations of the fundamental and pervasive intra-connectedness of the universe—e.g., perhaps greater realizations of dynamics such as quantum entanglement or other forms of "non-locality" more generally.
3. Something like a multiverse could exist as a robust realization of CAP as a source of "cosmic diversity" and endless forms of relations.
4. Some physical "laws" and/or constants may in some sense be mutable or vary in space and time,[26] in part because mutable laws and mutable constants could give rise to more novel relations and actions and creative evolutionary paths, increasing diversity and creativity over time.

8.7.2 *Practical Ethical Challenges*

When it comes to making practical decisions, either personally or collectively, that is when the ethical rubber hits the road and when many general principles can become insufficient or even counterproductive. The obvious difficult question here is how the connection-action principle could help with more day-to-day practical ethical questions, as well as more metaethical guidance that might help inform such deliberations.

Perhaps the biggest and broadest challenge of applying CAP to practical ethics is that even if the universe is about realizing its property of connectedness as action, can human beings or intelligent beings more generally really identify with that or use it in any sensible way? And further, why *should* we? The universe or multiverse may be the biggest, most compelling absolute we are aware of (perhaps other than suggestions of "spirit" or God), but it does not necessarily follow that we must then use the universe to obligate ourselves to something like CAP. This is what the naturalistic fallacy and the "is–ought" distinction properly sensitize us to. However, as noted previously, the kind of reasoning that takes us from a perceived reality of

[26] Some prominent physicists such as Paul Dirac, Richard Feynman, and John Wheeler have explicitly questioned the immutability of physical laws, and the idea has been extended by philosophers and physicists who suggest such a consideration is needed in physics to help address certain problems in physics and cosmology (Unger and Smolin 2015). Indeed, Webb et al. (2001) have interpreted their astrophysical measurements to suggest that the fine-structure constant varies slightly or "evolves." However, while Carroll (2010a, b) notes that some changing constants (probably *slow* changing) could be accommodated by existing theories, he, like others, thinks the fine-structure constant is effectively constant (at least for now—based on reasonable interpretations of the present evidence). We also previously touched on the idea that something like "cosmological natural selection" could give rise to different constants and different laws in different universes. Another example could be that the second law of thermodynamics may not really be a fixed physical law or a truly physical law in the deeper sense—or if it is, it will be overcome at cosmic scales in the same way life and intelligence "overcome" it now at local space- and timescales. Bloom (2012) suggests that the second law of thermodynamics is not really a physical law in the sense that we normally think about physical laws.

the universe to more value-oriented judgments based on that reality could simply be a choice that sufficiently aware beings can make if it is carefully reasoned.

While there might be some aspects of CAP that are consistent with human predispositions, presumably there are many that are not. For example, human beings very often value simplicity and a lack of diversity and a lack of action in many cases. If evolution can explain these predispositions, maybe we do not need more explanation or further basis for human values—we simply accept evolutionary values and make messy practical choices that attempt to create relatively stable, balanced, and rewarding lives. Maybe that is all that can and should be attempted in practical ethics and perhaps in our worldviews more generally. This kind of philosophical pragmatism is arguably consistent with scientific minimalism.

And even if we were to accept something like the connection-action principle as a way to assess practical ethical questions, how would it work in terms of the details for any given ethical challenge? For example, how would we weigh or measure the best "action" in any situation? Is it the *amount* of action, or *kind* of action, or both, or something else? And, how could it be assessed with satisfying precision, consistency, and credibility? While there may be ways to ultimately attempt this kind of assessment, it is also reasonable to suppose that it will be essentially impossible to make these assessments with enough consistency or credibility to satisfy many. There would presumably be many open-ended and highly contradictory value judgments about how best to realize connectedness through relationships and action—and that raises a flag to say the least.

Nevertheless, the idea of "value measurement" (in this case "measuring" relationality, interaction, and/or intra-action), while possibly having some undesirable effects such as being used improperly (as "survival of the fittest" arguably has been), could, in theory, be carefully pursued and based on careful assessments of what constitutes degrees of relationality and action in a way that is cautious and thoughtful. As noted previously, this could be very complicated but may ultimately have a helpful basis in science to the extent that things like basic interaction and intra-action, diversity and complexity, etc., are sufficiently definable, measurable, and perhaps interrelatable. Such "measurements" could in principle include many complicated factors including a wide range of human emotions, perceptions, experiences, and desires.

8.7.3 Cosmic Destruction and Cosmic Perpetuation?

The fact that there is much destruction in the world—a primary form being the basic need for life to destroy and consume life—can be interpreted to go against, or at least be inconsistent with, the connection-action principle. However, we might consider that this is just how the universe has evolved in our little corner of the galaxy, and that it need not necessarily be this way everywhere. Or maybe cultural evolution can lead to significant reductions, if not an elimination of this kind of destruction.

Despite the powerful logic of natural selection, it might be justified to consider sufficiently different forms of it elsewhere, or perhaps it may not be necessary at all for the evolution of certain kinds of life and intelligence—or at least for sufficiently complex "self-organizing" and self-replicating systems elsewhere.[27] Natural selection may be completely transcended by many forms of cultural evolution, including the evolution of artificial intelligence, for which reduced destruction might be an organizing ethic. Artificial intelligence, or "super-intelligences" may be able to more easily transcend destructive tendencies since presumably such intelligence will not be controlled by biologically driven tendencies and will have a broad base of knowledge and experience to draw from.

We might also consider that in some sense the universe, or multiverse, is not capable of perfectly manifesting CAP, so that there are limitations to instantiating connectedness (e.g., limitations to complexity).[28] So, what we observe as destruction, including the second law of thermodynamics, may be partly indicative of some kind of cosmological or "universal" limitation. Related to that, we might also consider the common sense notion that there must be destruction in order to have creativity—what is often referred to as "creative destruction." This may be indicative of an inherent limitation to the universe. Conservation laws may require it. The connection-action worldview does not require an "optimal" universe per se.[29] We might also consider that it is part of our purpose as living, intelligent creatures to help the universe actualize itself to greater extents—we may be a means by which unnecessary destruction, or destruction more generally, can be minimized or maybe even eliminated entirely.

The problem of destruction is more forcefully realized by considering that in an oscillating universe model, much it is often suggested that no form of life or any kind of information could survive an event such as big crunch or big bounce. Such complete destruction would seem to oppose the idea that the universe trends toward ever-increasing actualizations of connectedness—particularly increasing organized complexity like life and intelligence—if ultimately it is to only be completely destroyed. We might deal with this concern by again considering the notion that the destruction of what is created may ultimately be necessary for more new creations to subsequently emerge. This may simply be a fundamental limitation to the universe or any given universe. The universe might just get to a point where it must "start over" or "recycle" itself because there are certain inherent limitations to the universe and its evolution. Perhaps this recycling is "good" in that it allows for fundamentally new creations to emerge, for new actualizations of the universe's nature to be realized.[30]

---

[27] See Lupisella (2013) for brief speculation regarding a biological non-Darwinian intelligence.

[28] For example, John Smart explores a model that "would predict the necessity of a branching tree of self-organizing complexity underlying our universe, and an abundance of very simple proto-universes coexisting in the multiverse with a comparatively tiny number of complex universes such as ours." (Smart 2009, 231)

[29] Freeman Dyson notes his brand of Socinian theology as one where "God is neither omniscient nor omnipotent. God learns and grows as the universe unfolds." (Dyson 1988, 119). And we note again a related suggestion by John Corliss that the evolution of the universe is a learning process (Corliss 1989).

[30] Stapledon, in his classic science fiction novel, *Star Maker* (1937), explores the theme of endless universal cycles as part of an "infinite spirit's" (the Star Maker) attempts at perfect, but failing, creative activity.

In a cosmic "heat death," or what is sometimes referred to as the "big freeze" or "big chill," it appears action would essentially come to an "end."[31] However, if there is a multiverse constituted by many universes, then the end of one universe may not be a significant problem for realizing CAP—especially if universes give birth to more universes before they themselves come to end. A multiverse may be a way to in some sense make up for any destruction there may be in one universe by not having that particular destruction in another universe. This relates also to the quantum multiverse or "many-worlds" interpretation of quantum mechanics in the sense that constantly branching universes are essentially alternative realities, perhaps in which destruction in one universal timeline is avoided in another.

Worldviews that suggest the intentional creation of endless universes as a fundamental ethic (e.g., consistent with ideas from Vidal 2014 and others touched on in Chap. 2) would be consistent with the connection-action principle. It may very well be that life and intelligence are a vehicle by which the nature of the universe can be robustly realized as we evolve and add to the creations of the universe. We may very well be a way for the universe to realize its greatest potential, perhaps for an eternity. So perhaps the big crunch will not be the end. Maybe it will be possible to somehow survive a big crunch? Could it be avoided? Could we send information "through" such an event? Perhaps there will be a way for intelligent beings to survive it. For example, could intelligent beings become so embedded in the fabric of the universe (perhaps as forms of information embedded in the universe) that a big crunch will just become a "rebooting" opportunity with some of the same, yet somewhat different, operating "software" in place? This is similar to ideas explored in Chap. 2 and could arguably be motivated by CAP.

It is impossible to predict what life and the universe will be like at such long-term future epochs of course, but one general possibility is that intelligence and the universe will become so interconnected, so *intra*-connected, that the system will become one universal "mind" or universal "consciousness" as some have speculated (again, as partly touched on in Chap. 2). Indeed, it may be that for intelligent beings to survive and evolve (which, again, would arguably be consistent with actualizing the universe's nature as long as other robust creations were not prevented) such a "merge" might not be as fanciful as it sounds. If such a cosmic merge between mind and matter occurred, consciousness would no longer be a subset of the universe, and so it could be less vulnerable to destruction at a big crunch or other cosmic critical events that might occur, including heat death. Instead, "we" would completely and truly be the universe in some new, unimaginable form, forever existing, forever creating.

---

[31] Related, Freeman Dyson was one of the first to explore the possibility of life and communication continuing forever even as the universe approached a potential heat death (Dyson 1979).

8.8 Summary of Chapter 8: The Connection-Action Principle

In this chapter, we moved much further along the relational spectrum in Fig. 1.1 and introduced the connection-action principle (CAP) that suggests the universe's property of connectedness is manifested as relations and actions, perhaps necessarily and in ever-increasing degrees for stronger versions of CAP. The connection-action principle is consistent with, if not a foundation for, a relational metaphysics (although as Fig. 1.1 shows, CAP extends beyond the relational framework a bit to signify the possibility that it may not be constrained solely within the kind of relational framework explored in this book). But CAP is arguably consistent with, if not an explanation for relational interpretations of relativity and quantum theory—including relational quantum mechanics and quantum field theory that posits seemingly ubiquitous, infinite, interrelated dynamic fields.

CAP may provide a possible explanation for the ongoing creativity of cosmic evolution, including possibly within the context of a multiverse. CAP is also consistent with, and may offer a conceptual foundation for a number of philosophical concepts such as process philosophy, organic unity, intrinsic value, and the principle of plenitude, as well as other scientific concepts and constructs such as physical "laws" (with potential mutability), time, and emerging informational views of the unvierse (where information is the expression of relationships).

We briefly revisited the psychology of philosophy, specifically heightening our sensitivity to the potentially aesthetic appeal of notions of connectedness, relationality, and diversity, which arguably have some evolutionary roots and could be misleading tendencies. That psychological sensitivity, along with more detailed practical ethical considerations, led us to raise questions about the pragmatic utility of CAP and whether value measurement in the context of CAP would really be possible or credible. Nevertheless, we briefly considered at least theoretical paths forward for such a possibility.

We touched on the idea that much of the destruction we see in the world seems inconsistent with CAP, but we explored some possible ways to address this, including the possibility that there are important limitations to CAP, and/or that the universe may be imperfectly, incompletely, or perhaps just differently, realizing connectedness, relationality, and action with respect to our conceptions of CAP. We considered the possibility that the universe may have to in some sense "re-boot" with new cycles to further realize CAP and/or that the multiverse may be a way for a fuller, richer realization of CAP.

Despite its speculative nature, it can be argued that the connection-action principle is in some sense still a minimal metaphysical claim. While CAP may not be particularly scientific in the sense of being clearly testable or falsifiable, it nevertheless does not invoke specific or necessary supernatural entities, nor does it make absolute or dogmatic claims about its ultimate specific actualizations in the world. The connection-action principle relies primarily on the suggestion that the property of connectedness is ultimately manifested as relationality and action.

References

Alonso-Serrano, A., Bastos, C., Bertolami, O., & Robles-Perez, S. (2013). Interacting universes and the cosmological constant. *Physical Letters B, 719*(1–3), 200–205.

Albert, D. (2012 March 23). On the origin of everything. In *The New York Times Sunday Book Review*.

Altman, A. (1984). Nozick's theory of value and its implications. *Southern Journal of Philosophy, XXII*(1).

Barad, K. (2007). *Quantum physics and the entanglement of matter and meaning*. Durham and London: Duke University Press.

Barrow, J. D., & Tipler, F. J. (1986). *The anthropic cosmological principle*. Oxford University Press.

Bloom, H. (2012). *The God problem: How a Godless universe creates*. New York: Prometheus Books.

Bohm, D., & Hiley, B. J. (1993). *Undivided universe: An ontological interpretation of quantum theory*. New York: Routledge.

Bohm, D. (1980). *Wholeness and the implicate order*. London: Routledge & Keegan Paul.

Bokulich, A., & Jaeger, G. (2010). *Philosophy of quantum information and entanglement*. Cambridge University Press.

Bostrom, N. (2003). Are you living in a simulation? *Philosophical Quarterly, 53*(211), 243–255.

Briceño, S. & Mumford, S. (2016). Relations all the way down? Against ontic structural realism. In A. Marmodoro & D. Yates (Eds.), *The Metaphysics of Relations* (pp. 198–217). Oxford University Press.

Carroll, S. (2010a). *From eternity to here: The quest for the ultimate theory of time*. New York: Dutton.

Carroll, S. (2010 October 18). The fine structure constant is probably constant. Preposterous Universe: http://www.preposterousuniverse.com/blog/2010/10/18/the-fine-structure-constant-is-probably-constant/.

Cleland, C. (1984). Space: An abstract system of non-supervenient relations. *Philosophical Studies, 46*, 19–40.

Corliss, J. (1989). The flow of energy, natural learning systems, and the creation of life on earth. *Acta Astronautica, 19*(11), 869.

Davies, P., Gregersen, N. (2010). *Information and the nature of reality: from physics to metaphysics*. Cambridge University Press.

Davies, P. (1999). *The fifth miracle: The search for the origin and meaning of life*. New York: Simon and Schuster.

Davies, P. (2009). Life, mind, and culture as fundamental properties of the universe. In S. J. Dick & M. L. Lupisella (Eds.), *Cosmos and culture: Cultural evolution in a cosmic context*. Washington, DC: NASA History Series. http://history.nasa.gov/SP-4802.pdf.

Deutsch, D. (2011). *The beginning of infinity: Explanations that transform the world*. London: Penguin Books.

Delahaye, J. P., & Vidal, C. (2018). Universal ethics: Organized complexity as an intrinsic value. In G. Y. Georgiev, C. F. Martinez, M. E. Price, & J. M. Smart (Eds.), *Evolution, development and complexity: Multiscale evolutionary models of complex adaptive systems*. Springer.

Dyson, F. (1988). *Infinite in all directions*. New York: Harper & Row.

Dyson, F. J. (1979). Time without end: Physics and biology in an open universe. *Reviews of Modern Physics, 51*(3).

Esfeld, M. (2016). The reality of relations: The case from quantum physics. In A. Marmodoro & D. Yates (Eds.), *The Metaphysics of Relations* (pp. 218–234). Oxford University Press.

Guth, A., & Steinhardt, P. (1984). *The inflationary universe*. Scientific American.

Guth, A. (1998). *The inflationary universe: The quest for a new theory of cosmic origins*. Perseus Books.

Hall, M. J. W., Dirk-André, D., & Wiseman, H. (2014). Quantum phenomena modeled by interactions between many classical worlds. *Physical Review X, 4*, 041013.

Handy, R. (1970). *The measurement of values*. St. Louis: Warren H. Green Inc.
Francis, H. *Self-organization of complex, intelligent systems: An action ontology for transdisciplinary integration* (ECCO Working Paper 2010–03). Retrieved from http://pespmc1.vub.ac.be/papers/ECCO-paradigm.pdf.
Heylighen, F., & Beigi, S. (2018). Mind outside brain: A radically non-dualist foundation for distributed cognition. In A. Clark, S. O. Palermos, & D. Pritchard (Eds.), *Socially extended epistemology*. Oxford University Press.
Kashlinsky, A., Atrio-Barandela, F., Kocevski, D., Ebeling, H. (2008). A measurement of large-scale peculiar velocities of clusters of galaxies: Results and cosmological implications. *The Astrophysical Journal Letters, 686*(2).
Kauffman, S. (1995). *At home in the universe*. New York: Oxford University Press.
Khoury, J., Ovrut, B. A., Steinhardt, P. J., & Turok, N. (2001). Ekpyrotic universe: Colliding branes and the origin of the hot big bang. *Physical Review D, 64*.
Krauss, L. (2012). *A universe from nothing: Why is there something rather than nothing?*. New York: Free Press.
Lemos, R. (1995). *The nature of value: Axiological investigations*. University Press of Florida.
Leopold, A. (1966). *A sand county almanac*. New York: Oxford University Press.
Linde, A. (1986). Eternally existing self-reproducing chaotic inflationary universe. *Physical Letters B, 175*(4), 395–400.
Linde, A. (1994). The self-reproducing inflationary universe. *Scientific American*.
Lloyd, S. (2006). *Programming the universe: A quantum computer scientist takes on the cosmos*. New York: Random House.
Lovejoy, A. (1936). *The great chain of being*. Harvard University Press.
Lupisella, M. (2013). Caring capacity and cosmocultural evolution: Potential mechanisms for advanced altruism. In D. A. Vakoch (Ed.), *Extraterrestrial altruism*. Berlin: Springer.
Mackenzie, R., Shanks, T., Bremer, M., Yan-Chuan, C., Gunawardhana, M., András Kovács, A., et al. (2017). Evidence against a supervoid causing the CMB cold spot. *Monthly Notices of the Royal Astronomical Society, 470*(2), 2328–2338.
MacNiven, D. (1993). *Creative morality*. New York: Routledge.
Marshall, A. (1993). Ethics and the extraterrestrial environment. *Journal of Applied Philosophy, 10*(2).
Marshall, A. (2002). *The unity of nature: Wholeness and disintegration in ecology and science*. London: Imperial College Press.
Mermin, N. D. (1998). What is quantum mechanics trying to tell us? *American Journal of Physics, 66*, 753–767.
Mermin, N. D. (2017). Why QBism is not the copenhagen interpretation and what John Bell might have thought of it. In R. Bertlmann & A. Zeilinger (Eds.), *Quantum [Un]speakables II*. The frontiers collection. Springer International.
Mersini-Houghton, L., Holman, R. (2009). 'Tilting' the universe with the landscape multiverse: the 'Dark' flow. *Journal of Cosmology and Astroparticle Physics*.
Moore, G. E. (1903). *Principia Ethica*. Cambridge University Press.
Murphy, N., & Ellis, G. (1996). *On the moral nature of the universe; theology, cosmology, and ethics*. Minneapolis: Fortress Press.
Nozick, R. (1981). *Philosophical explanations*. Cambridge: Harvard University Press.
Oliver, H. (1981). *A relational metaphysics*. Boston: Matinus Nujhoff Publishers.
Quante, M., & Engelhard, K., Eds. (2018). *Handbook of Potentiality*. Netherlands: Springer Netherlands.
Rescher, N. (1996). *Process metaphysics: An introduction to process philosophy*. State University of New York Press.
Rescher, N. (1998). *Complexity: A philosophical overview*. New Brunswick: Transaction Publishers.
Rescher, N. (2006). *Metaphysics: The Key Issues from a Realistic Perspective*. Amherst, NY: Prometheus Books.
Rollin, B. (1981). *Animal rights and human morality*. Buffalo: Prometheus Books.

Rolston, H. I. I. I. (2012). *A new environmental ethics: The next millennium for life on earth*. New York: Routledge.
Roltson, H. III. (1986). The preservation of natural value in the solar system. In E. C. Hargrove (Ed.), *Beyond spaceship earth: Environmental ethics and the solar system*. San Francisco: Sierra Club Books.
Sciama, D. W. (1959). *The unity of the universe*. Garden City: Doubleday & Company.
Singer, P. (1981). *The expanding circle: Ethics and sociobiology*. New York: Farrar, Straus & Giroux.
Smart, J. (2009). Evo devo universe? A framework for speculations on cosmic culture. In S. J. Dick & M. Lupisella (Eds.), *Cosmos and culture: Cultural evolution in a cosmic context*. Washington, DC: NASA History Series.
Smith, K. (2014). Manifest complexity: A foundational ethic for astrobiology? *Space Policy, 30,* 209–221.
Smolin, L. (1997). *The life of the cosmos*. New York: Oxford University Press.
Stapledon, O. (1937). *Star maker*.
Steinhardt, P. J., & Turok, N. (2002a). A cyclic model of the universe. *Science, 296,* 1436.
Steinhardt, P. J., & Turok, N. (2002b). Cosmic evolution in cyclic universe. *Physical Review D, 65,* 126003.
Strand, M. (1968). "Keeping Things Whole" in Reasons For Moving, New York: Atheneum. Originally titled "A Reason For Moving", published in 1964 in Sleeping With One Eye Open, Stone Wall Press (Iowa City). Also included in a more recent 2014 volume, Collected Poems, New York: Alfred E. Knopf.
Tegmark, M. (2003). Parallel universes. Not just a staple of science fiction, other universes are a direct implication of cosmological observations. *Scientific American, 288*(5), 40–51.
Timpson, C. G. (2010). Information, immaterialism, instrumentalism: Old and new in quantum information. In A. Bokulich & G. Jaeger (Eds.), *Philosophy of quantum Information and entanglement*. Cambridge University Press.
Turchin, V. F. (1977). *The phenomenon of science: A cybernetic approach to human evolution* (Translated by B. Frentz). Columbia University.
Turchin, V. F. (1993). The cybernetic ontology of action. *Kybernetes, 22*(2), 10–30.
Unger, R. M., & Smolin, L. (2015). *The singular universe and the reality of time*. Cambridge University Press.
Volk, T. (2017). *Quarks to culture: How we came to be*. Columbia University Press.
Von Neumann, J. (1932). *The mathematical foundations of quantum mechanics*. (R. T. Beyer, Trans.). Princeton Univ. Press. (1996 edition).
Webb, J. K., Murphy, M. T., Flambaum, V. V., Dzuba, V. A., Barrow, J. D., Churchill, C. W., et al. (2001). Further evidence for cosmological evolution of the fine structure constant. *Physical Review Letters, 87,* 091301.
Webb, S. (2004). *Out of this world: Colliding universes, branes, strings, and other wild ideas of modern physics*. New York: Copernicus Books.
Wheeler, J. A. (1983). Law without law. In J. A. Wheeler & W. H. Zurek (Eds.), *Quantum theory and measurement*. Princeton University Press.
Wheeler, J. A. (1990). Information, physics, quantum: The search for links. In W. Zurek (Ed.), *Complexity, entropy, and the physics of information*. Redwood City: Addison-Wesley.
Wolfram, S. (2002). *A new kind of science*. Wolfram Media, Inc.
Zenzen M. J., & Hammer, L. (1978). Value measurement and existential wholeness. *The Journal of Value Inquiry, XII*(2).

Chapter 9
Meaning and Ethics

I think a future flight should include a poet, a priest and a philosopher…we might get a much better idea of what we saw.
—Michael Collins, Apollo 11Astronaut

A number of contemporary cultural phenomena express hunger for a new cosmopolis. The religious and quasi-religious speculations of scientists, and the great popularity of their books, suggests a hunger to relate our burgeoning knowledge of the cosmos to the pursuit of human meaning…
—Nancy Murphy and George Ellis (1996, p. 2)

9.1 Background and Overview

The meaning of meaning can be slippery. It can be broad, vague, diverse, and very subjective—even more so than ethics which often tries to seek more practical common ground than meaning-seeking. The intent of this book is not to cover the subject of meaning or worldviews in general, for which there is already much academic and popular literature (e.g., Aerts et al. 1994, 1999; Babbage and Ronan 2000; Wilson 2002; Shermer 2002; Koltko-Rivera 2004; Gershenson et al. 2007; Bulbulia et al. 2008; Johnson et al. 2011; Henriques 2011; Vidal 2012, 2014; Nilsson 2013; Hedlund-de Witt et al. 2014; Saucier 2013; Saucier et al. 2015). Instead, we are exploring a few "cosmological theories of value" and the associated "cosmological worldviews" that may have the potential to inform a sense of meaning in whatever way might make sense to the reader. In this broad and admittedly loose "meaning-making" context, we will briefly revisit some motivational psychology and then explore a practical ethical challenge regarding potential life on Mars.

In Part 1, we covered the science of relativity, quantum theory, and cosmic evolution, along with some key philosophy such as value theory and process philosophy to help establish an overall relational framework which then acted as a context and foundation for Part 2 that explored three cosmological theories of value or worldviews: cosmological reverence, cosmocultural evolution, and the connection-action principle. As explored previously, these three perspectives can be interpreted to support varying forms and degrees of partially overlapping cosmological theories of

M. Lupisella, *Cosmological Theories of Value*, Space and Society,
https://doi.org/10.1007/978-3-030-25339-4_9

value, which can then be used to help provide (a) some sense of meaning in a cosmic context, along with (b) the idea of a morally creative cosmos, and (c) some forms of cosmic ethics, or perhaps cosmo*centric* ethics.

To help set the stage a bit further, let us consider these somewhat cryptic excerpts from a 1967 essay from James Feibleman, on "Normative Cosmic Ethics"—some of which resonates with what we have explored so far:

> It might just be possible, then, to continue our description of cosmic ethics by adding a normative section. In terms of the organization of matter, a galaxy can become more what it is. Self-realization though unconscious in such cases can still be genuine. And if it proceeds with incredible slowness it may be no less effective for that; a product of law and with a direction modified by chance encounters.[1]

Feibleman further writes (admittedly chauvinistically):

> Men are more interested in doing than they are in being, because it seems to them that their only hope lies through activity. Thus their interest becomes narrowed to the issue of activity in terms of human problems. But it may be that the larger issue contains information and values which could be applied to the smaller, that there are cosmic considerations which could be brought to bear successfully upon human consideration. Man may not be necessary to the world but it is necessary to him. However, although he is a small part of the world he is an authentic ingredient and cannot legitimately be considered apart from it. For instance, if all material things are interrelated, and if the fate of any thing depends, as it seems to do, upon that of all others, then if all men are brothers, all animals are first cousins and all lower organisms second cousins and all material things cousins of more remote degree but cousins. Men, then, are cousins to material things.
>
> In terms of human welfare, because of relevancy, nothing in the environment can be neglected. Completeness is an all-inclusive requirement. Cosmic ethics, then, is simply a way of saying that what brings things together is good and what sets them apart is bad. (Bringing things together in the wrong order, however, is a way of setting them apart.)[2]

9.2 Psychology Revisited: Turning the Lens of Evolutionary Psychology

In Chap. 3, we briefly touched on the psychology of philosophy, noting how some psychological predispositions can find affinity with certain kinds of philosophical orientations and philosophical musing in general. Feibleman's excerpt above arguably simultaneously attempts to transcend human psychology to cosmic scales, while at the same time using appealing human psychological constructs such as family metaphors (e.g., humans as "cousins" to matter). Here we will further explore the "psychology of philosophy," particularly as it might apply to the connection-action principle. For example, we previously noted Robert Nozick's observation that the pursuit of an objective basis for values implies the value judgment that it is better

[1] Feibleman (1967), p. 192.

[2] Feibleman (1967), p. 193.

to have objective values as opposed to subjective values. This kind of perspective could be influenced heavily by particular psychological tendencies or predispositions for which evolutionary psychology may be able to shed light.[3]

For example, if group members can be convinced that there is an objective basis for values, then those values may carry more force and take hold better within the group. "Objective" values may be perceived to have a source that is not restricted solely to what other group members think or want for themselves. Instead, such objective values could be perceived to be associated with something beyond the interests of other group members, perhaps with something much more powerful or all-encompassing—something more "deeply" true and highly consequential. And if those values help a group to be tightly bound, to act cohesively (and aggressively toward out-groups), to act as good resource managers, and to generally learn better about what is independently real about the environment, then such tendencies for seeking objectivity could persist in subsequent generations of groups via group-level selection.

Similarly, and more specifically with respect to the connection-action principle, certain psychological architectures may be predisposed to be highly sensitive to connections and relationships in order to navigate and survive in a complex natural and social world. A sensitivity to relationships, or relationality more generally, can help individuals and groups understand the causal world better. A sensitivity to connectedness and relationships in one's group can make an individual a good group member and help increase personal and group fitness. Emphasizing the "connectedness" of everything could have group cohesion value.

A sensitivity to relationships is consistent with how the idea of a "relational self" (noted in Chap. 2) can stem from child rearing (Barlow 1997). Related to that, love, and how it may motivate familial and group connectedness, may be appealing because of its power to motivate a sense of belonging and commitment to children and one's group. A general sensitivity to action in the world, perhaps in the form of being "active," e.g., in motion, producing, accomplishing tasks, etc., presumably can also increase the fitness of individuals and groups.

Diversity may be appealing because it may help a group be prepared for different challenges and perhaps also help "outlier" individuals promote values that would allow them to be better tolerated by a group.[4] Extreme diversity, perhaps even in the form of the principle of plenitude, may also have a certain appeal in the sense that it allows for all things, all actions, all possibilities to be realized—perhaps creating the impression that constraints can be overcome. Related to that, putting ourselves in a privileged position of being powerful creators, co-creators and arbiters of value and reality (e.g., per cosmocultural evolution) can also be psychologically comforting

[3]Related, Rescher (1985) makes a useful distinction between epistemic values and other kinds of values, where epistemic values are values that help inform how we learn about the world. In this context, seeking an objective basis for values could be considered an epistemic value—including the possibility of having evolved by natural selection.

[4]Haidt (2012) discusses how group selection can produce different levels of psychological predispositions toward diversity in order to satisfy certain group functions and also enhance individual fitness.

and empowering in the sense that it confers great privilege and power to ourselves—a tendency for which there is an obvious problematic history.

Highly active minds, or unusually hyperactive minds, may be drawn to action (particularly as action relates to productivity), diversity, creativity, and complexity and may find those notions appealing because it can keep people active and engaged. This kind of appeal would merely be a result of how such brains work (perhaps as a result of selective pressures), not necessarily because there is anything truly objective or independently valuable about such tendencies. Indeed, perhaps the simple process of natural selection can predispose us to see more connectedness and complexity than there really is—and to see patterns, meaning, and purpose where there may otherwise be none. We should not underestimate how much the simple mechanism of natural selection can produce so much complexity, including unnecessarily complex mental states and behaviors.

So, perhaps behind much philosophical musings and meaning-making are important but sometimes misleading psychological motivations, particularly evolutionary motivations driven by individual and group selection—at least as "by-products" of such selective pressures if not direct fitness-enhancing pressures. But exploring these kinds of psychological sensitivities and factors is not meant to fundamentally undermine philosophical pursuits or philosophical psychological predispositions. It is meant more to heighten our sensitivity and encourage a certain amount of humility and open-minded caution in our philosophical explorations—particularly those of a highly speculative and sweeping nature such as those involving the whole universe.

9.3 A Moral Cosmos?

> To be a moral human being is to pay, be obliged to pay, certain kinds of attention.
> —Susan Sontag [5]

9.3.1 Cosmological Reverence

Cosmological reverence is the weakest of the three cosmological theories of value in terms of consequence and value, but nevertheless ascribes significant value to a creative cosmos that has given rise to beings such as ourselves who are in turn highly

[5] Sontag, Susan. 2007, *At the Same Time: Essays and Speeches*. Eds. Paolo Dilonardo and Anne Jump. New York: Farrar, Straus, Giroux. Original Essay, "The Truth of Fiction Evokes Our Common Humanity", given by Susan Sontag April 7, 2004 at the Los Angeles Public Library upon receiving the library's Literary Award. Published in the Los Angeles Times, December 29, 2004. Maria Popova helps us pay attention with her article here about Sontag's emphasis on the importance of paying attention: https://www.brainpickings.org/2015/03/30/susan-sontag-writing-storytelling-at-the-same-time/.

creative. This kind of view sees great value in the apparent scientific fact that we and all living and non-living things are fundamentally related to, and are products of, an astonishingly long and complex history cosmic evolution. Cosmic evolution, and its process-oriented relational nature, is revered in cosmological reverence.

Cosmological reverence views ourselves and other beings in essentially a "one-way" relationship with the universe in that we are products of cosmic evolution, and so the universe is significant for us, but we do not have significance for the universe—we are merely fortunate beneficiaries. Nevertheless, this recognition and other kinds of relationships in our dynamic and creative universe are deeply valued and celebrated. Cosmological reverence can provide comfort, inspiration, and awe for many people and can enrich the human experience. For some, it may be possible for cosmological reverence to also provide some comfort and perhaps even guidance for how we view and treat our environment and other beings.

9.3.2 Cosmocultural Evolution

Cosmocultural evolution goes a good bit further than cosmological reverence by suggesting that we have a "two-way" relationship with the universe in the sense that not only is the universe significant for intelligent, valuing, cultural agents like ourselves, but we and any other intelligent cultural beings (where cultural implies something like "collective valuing") throughout the universe have significance for the universe—in part because we help the universe bootstrap itself into the realm of value, meaning, purpose, and ethics.[6] Cosmocultural evolution points us to the idea that sufficiently intelligent beings can see themselves as a form of consciousness for the universe, as creators and arbiters of value and meaning—as a way for the universe to *value* itself.

With this kind of context in mind, we can choose to make the well-being of sentient beings a priority, to find meaning and purpose in each other, not just because we may be "on our own" as a species in a universe that may not care about us in the way we might wish, but also because in caring for each other, in loving each other, we can bring that caring and love to the universe itself. We can help love emerge in the cosmos and make the universe a moral entity with love as a kind of "force." And if culturally evolving sentient beings are ultimately significant for the universe as a whole, physically or metaphysically, then loving and caring for each other helps realize a kind of significance we have for the universe.

We can see ourselves then as arbiters, creators, and co-creators of open-ended, ever-evolving forms of thought, reasoning, value, meaning, purpose, and creativity—not just for ourselves, but for the whole of the cosmos. We can choose to carefully value, act, and create wisely with those values—including helping the universe to become an increasingly morally creative cosmos.

---

[6] Jonas (1984) suggests that preserving humanity is a way to preserve the presence of ethics in the universe.

9.3.3 Connection-Action Principle

The connection-action principle goes much further than cosmological reverence and cosmocultural evolution. The idea that the universe may be realizing its fundamental property of connectedness through relationships and actions (perhaps necessarily so and in ever-increasing degrees) might help inform our perceptions of value and meaning of any ethical and moral sentiments by seeking out and creating diversity, novelty, and perhaps high degrees of complexity, dynamism, and creativity in general. This kind of view, while certainly problematic for a number of reasons touched on previously, could still help guide our ethical perspectives by recognizing a broader base of value beyond the solely subjective value of our individual and collective predispositions and deliberations—most of which appear to be largely influenced by Darwinian evolution.

The connection-action principle suggests not only that we can recognize and revere a highly connected, relational, dynamic, and diverse universe, but that we can also encourage those characteristics in our contemplations and actions and act as co-creators of a morally creative cosmos—perhaps with other extraterrestrial beings who would also be seen as robust actualizations of the nature of the universe and hence worthy of respect.[7] We might generalize even further, noting that some ethical interpretations of the connection-action principle could suggest that our connectedness and relationships to to each other (human or otherwise), suggest the importance of taking action on behalf of each other.

9.4 Cosmic Ethics: A Martian Scenario[8]

> ...stark and silent...were the Martians—dead!—slain by the Humans against which their systems were unprepared...slain, after all the Human's devices had worked, by the blind foreigners that had landed upon their world. Yet across the gulfs of space, minds that were to Humans as Humans were to the Martians that perished, intellects vast and cool and unsympathetic, regarded this earth with contempt, and slowly and surely drew their plans against us—we who had killed another. (Adapted from *War of the Worlds* by H.G. Wells)[9]

Having suggested foundations of value theory that relate to cosmological considerations, we can now apply those views to an unusual ethical challenge. As touched on previously and as shown in Figs. 1.1 and 6.1, cosmocentric ethics can relate to

[7] Related to the connection-action principle and ideas of a moral universe, Burkhart (2004) points out how Native American thought often invokes principles (or "ways of being") he calls the "meaning-shaping principle of action" and "the moral universe principle" which taken together can be interpreted to suggest that through our actions we can help realize and enhance a kind of universal morality that either already exists or that we can help bring to the universe (similar to what is suggested by cosmocultural evolution).

[8] This section is taken largely from the paper "Do We Need a Cosmocentric Ethic?" presented by M. Lupisella and J. Logsdon at the 1997 International Astronautical Congress, Turin, Italy.

[9] Adapted from *War of the Worlds* by H. G. Wells. Originally in Lupisella and Logsdon (1997).

and possibly be "derived" from the worldviews explored in this book. Here we will emphasize stronger versions of cosmocentric ethics, based primarily, but not solely, on the connection-action principle.

The connection-action principle arguably provides a foundational metaphysics as a way to more strongly support the idea of a cosmocentric ethic since it provides something foundational and consequential in the form of a proposed fundamental characteristic of the universe, i.e., connectedness and its implied "derivatives" such as action, diversity, and complexity. Stronger versions of cosmocentric ethics could be characterized with criteria such as (1) establishing the universe as a priority in a value system, (2) appealing to something characteristic of the universe, which might then (3) provide a justification of value, perhaps intrinsic value, and perhaps (4) allow for some kind of assessment of degrees of value (Lupisella and Logsdon 1997, Lupisella 2019).

At first glance, talk of a cosmocentric ethic might seem paradoxical. How can an ethical view be centered or focused on "all that is"? From egocentrism to geocentrism, we are able to center, focus, and prioritize value because there is often some other, generally broader frame of reference which is relatively devalued. Nevertheless, we can explore the possibility that such an ethic may be helpful in dealing with value-based questions of many kinds, including those involving extraterrestrial issues such as interaction with extraterrestrial life.

Here we will explore the importance of a practical policy issue facing us today regarding the exploration of our solar system,[10] particularly with respect to the first human presence on Mars. It is possible that we could adversely affect or cause the extinction of indigenous Martian life, intentionally or otherwise, by landing humans on Mars and eventually settling the planet.

Ethical and policy questions regarding possible Martian life can be categorized as pre-detection, post-detection, and longer term. Pre-detection issues involve those that are relevant for planning before a discovery is made. Post-detection issues concern how and what we decide to do after a discovery. This kind of planning is similar to work being done within the SETI community where protocols have been established regarding what to do after a signal has been received, if and how to go about communicating with ETI, etc. (Dick 1995; IAA SETI Permanent Committee 2010,[11] Vakoch 2011; Dick 2015, 2018). Pre-detection and post-detection questions can be further categorized as robotic vs. human exploration. There are also longer-term

---

[10] See Race (2013) for a treatment of decision-making factors and broader societal considerations regarding searches for extraterrestrial life.

[11] See the International Academy of Astronautics SETI Permanent Committee Web site (http://iaaweb.org/content/view/396/554/) for: (a) a description of a proposal called the "Rio Scale" for assessing the significance of a SETI detection, (b) a position paper outlining a process for examining the possibility of transmitting from earth, and (c) The San Marino Scale for quantifying the potential hazard of transmitting signals to potential extraterrestrial civilizations. See also Almar and Schuch (2007) for a paper on the San Marino Scale.

considerations involving how we might see Mars and putative Martian life after the scientific novelty and utility is sufficiently exhausted.[12]

9.4.1 Pre-detection Issues

Pre-detection planning regarding robotic missions has been addressed broadly in the form of the 1967 Outer Space Treaty and more specifically in the form of contamination prevention measures implemented by space-faring nations. However, the absence of rigorous international enforcement mechanisms may have allowed unacceptable contamination to occur in the past and may allow it to continue in the future. As more nations become space-faring, more effective international mechanisms for enforcing contamination regulations may be needed.[13]

Pre-detection planning regarding in situ human space exploration is beginning to be addressed—particularly as it relates to potential life-bearing planets such as Mars.[14] To what extent will there be contamination of the environment?[15] How will such contamination jeopardize or mask the existence of indigenous life-forms?[16] Will contamination due to a human presence be local or global?[17] Will it penetrate subsurface environments?[18] If there is a concern for global contamination, what will be the criteria for determining the biological status of the entire planet—or,

[12] For recent wide-ranging treatments of ethical issues associated with space exploration and astrobiology, see Schwartz and Milligan (2016), Impey et al. (2013), and Smith and Mariscal (2020). Brian Green (2020) makes a case that there are some convergences in a variety of ethical views regarding space activities and astrobiology (e.g. the importance of protecting extraterrestrial life).

[13] See Ehrenfreund et al. (2013) for a COSPAR Report on "Developing a Responsible Environmental Regime for Celestial Bodies." For a historical legal analysis, see Cypser (1993).

[14] The Moon is widely regarded to be a dead body, despite apparent prospects for water ice. See the International Council for Science Committee on Space Research (COSPAR) Planetary Protection Policy (2015) that includes Principles and Guidelines for Human Missions to Mars (https://cosparhq.cnes.fr/sites/default/files/ppp_article_linked_to_ppp_webpage.pdf). See NASA planetary protection reports that are beginning to address planetary protection for human missions (https://sma.nasa.gov/sma-disciplines/planetary-protection)

[15] McKay and Davis (1989) note that a human presence on Mars will result in contamination. They write: "It may be assumed, a priori, that all space suits and habitats will leak."

[16] For an early discussion on potential ecological impacts see Marshall (1993). For a discussion of pathogen/host co-evolutionary dependency and other factors suggesting contamination issues may not be important, see Zubrin's and Wagner's, *The Case For Mars,* (1996, 134–135). For a brief response, see Lupisella (1997), "The Rights of Martians?" in *Space Policy* Vol. 13, No. 2, p. 93.

[17] See for example, Glavin et al. (2004) for a possible strategy to learn more about contamination resulting from human space missions by studying the Apollo lunar landing sites to "baseline" the contamination at those locations before further activity at those sites introduces more contamination that might complicate our understanding of the original contamination from the Apollo missions.

[18] McKay and Davis suggest that contamination will probably not be global due to oxidation and ultraviolet radiation. They also write: "niches at considerable depths below the permafrost are probably safe and would not be affected by a human base" (1989, 198). See also Rucker (2013) for a NASA Mars drilling system study.

otherwise, just the landing site area or larger contamination footprint beyond the immediate landing site? Can we adequately extrapolate, either globally or locally, from a few strategic missions? If not, how many missions of what kind will be needed to obtain confidence about the status of life on Mars and any harmful effects our presence might have?[19]

Is the proven tenacity of life on Earth evidence that if we do not find life on Mars with a few missions, that it is indeed a dead planet? Or is it the other way around—that the ability of terrestrial life to survive in extreme environments should warrant significant caution, perhaps in the form of many precursor life-detection missions before sending humans? Can we have confidence in extrapolating from terrestrial biology if we essentially have only one data point—one world with one kind of life, apparently from a single common origin?[20] Our ability to have confidence from terrestrial extrapolation could be a key factor in assessing how we might interact with other worlds (Lupisella 2004, Cleland 2019).

If we conclude that many precursor missions are necessary before landing humans, how many should there be and will there be enough time and patience to conduct those robotic missions? Will we want to intentionally implant life on Mars during a robotic phase of Mars exploration (Haynes and McKay 1990)? Will we wait to land humans if it is thought human surface missions could compromise indigenous life and conduct surface operations from Mars orbit via low-latency teleoperations or "telepresence"? (Lupisella and Race 2017)

If it is thought that Martian life could be compromised by a human presence on Mars, how will the value of that potential life compare to the value of landing humans? As previously noted, some of the basic science is indeed already being pursued to help assess whether there might be life on Mars and whether our presence could compromise it or the search for such life. Looking for water, subsurface or otherwise, is a recognized key step, as is the detection of organic molecules—both of which are being actively pursued with robotic missions. Mars sample return missions are being planned, including plans for obtaining many samples from a fairly large area.

Operational guidelines for activities that could jeopardize indigenous ecosystems while humans are present on the surface would also be helpful (Race and Randolph 2002) and are increasingly being considered. Contamination control measures are a part of this, but there are also issues such as (a) establishing surveillance procedures before entering a potential biofriendly area (what are sometimes called "special regions"), (b) guidelines for crew and asset movements in certain areas, and (c) procedures for digging, drilling, releasing waste, and dealing with rocket exhaust are just a few areas for consideration. And again, much work is underway to understand these issues (Bobskill and Lupisella 2014; NASA & COSPAR Workshop Reports: Race

[19]These questions and others are explored further and are represented in a high-level decision tree in Lupisella (1999) "Ensuring the Scientific Integrity of Possible Martian Life." Paper IAA-99-IAA.13.1.08 presented at the International Astronautical Federation Congress. American Institute of Aeronautics and Astronautics, Amsterdam.

[20]Davies et al. (2009) explore the possibility of a "shadow biosphere" on Earth that may be quite different from life as we know it so far, including a separate origin.

et al. 2015, 2018, 2019, and Kminek et al. 2016).[21] Such guidelines for pre-detection activities while humans are on the surface may help preserve key environments where life could exist, undetected—or possibly *undetectable*. Emphasizing minimally invasive procedures may be one such guideline, e.g., humans operating robots in a special region from a close but safe distance (e.g., via low-latency teleoperations touched on prior).

9.4.2 Post-detection Issues

A discovery of indigenous life by a robotic asset may not present any significant difficulties if we take the proper contamination precautions and if we are willing to take the time needed after the discovery to make careful policy decisions about how to proceed—which will be driven largely by the scientific and programmatic circumstances and how much that life is valued. It may be prudent to consider some of these robotic post-detection issues now in order to prepare ourselves to whatever extent is appropriate (Lupisella 1999; Race and Randolph 2002; Ehrenfreund et al. 2013; Dick 2015, 2018).

Will we send humans immediately to a site that has evidence of extant life? Will we opt for an immediate sample return of those life-forms? Or, might we take a more conservative approach and study that life via robotic explorers so as to not disturb the immediate discovery site? If we choose robotic exploration, will it be of a remote nature, say from a low orbit or Phobos or Deimos via low-latency telerobotics (Lupisella et al. 2017), or will we land one or many vehicles at the immediate site as soon as possible? If it is possible to sufficiently control our contamination, and/or if there is sufficient reason to think our contamination will not cause adverse impacts to indigenous life (potentially two very challenging questions), then we may choose to land very close to the site of interest to permit direct human exploration.

It may be prudent to address these kinds of questions now to whatever extent we can while we are likely to have time after a robotic discovery is made. We may not have that luxury if humans make the first discovery while on the surface. There will likely be momentum, political and scientific, which could be hard to curtail, especially once humans are there.[22] Most importantly, with humans on the scene, it will be prudent to at least establish in advance some decision-making mechanisms, presumably of an international nature, to deal with post-detection activities. For example, if and when such a discovery is made, should an astronaut take a sample immediately? Should crew members leave the immediate site and do remote analysis before disturbing the site any further? Should the entire crew leave the surface entirely? The details

[21] McKay and Davis note several sources of environmental impacts due to a human base that should be considered, including mechanical disturbances, life support system leakage, airborne pollution, and "seemingly innocuous perturbations" like water, heat, light, etc. (McKay and Davis 1989, 198).

[22] McKay and Davis write: "It is arguable that once humans land on Mars, attempts to maintain a strict policy of preventing the introduction of Earth life into the martian environment will become moot" (1989, 197).

of any discovery will obviously be critical for effective decision-making regarding these questions.

Longer-term, once we have studied Martian life and understand it well, once we have exploited its scientific value to our satisfaction, what then? Does Mars belong to the Martians as Carl Sagan has suggested, “even if they are only microbes” (Sagan 1980, 108)? What, if any, moral consideration do we afford Martian life or extraterrestrial life in general? What might make extraterrestrial organisms worthy of our ethical consideration, and how do we go about addressing such a question given the apparent limitations of prevailing ethical views and lack of knowledge of putative Martian life? (Wilson and Cleland 2015). Does Mars belong to humanity? Does it belong to humanity *and* indigenous life? Does it “belong” to anyone at all?

9.4.3 The Potential Value of Martian Life

Underlying many of the previous challenges are questions of value. How much do we value the preservation of “primitive” extraterrestrial life such as microorganisms and why?[23] Certainly, there is instrumental value, or more specifically, scientific value. Masking the existence of such life and/or destroying it beyond recognition would be a scientific loss of immense proportion. Biology is desperate for a “second data point” (another kind of life), and there are many important questions that need consideration if we are to ensure the benefits associated with this scientific value.

However, it is not clear that scientific value will be enough to warrant the kind of conservative approach that may be needed to ensure the preservation of possible indigenous extraterrestrial life, thereby realizing that scientific value. As history has painfully demonstrated, the momentum of accomplishing certain goals, particularly big goals, often overshadows contemplation of consequences and any associated policy needs. The exploration and exploitation of the Americas are a poignant example of the harm our species is capable of—in that case, it was both the unintended consequences and intentional use of microorganisms for genocide (e.g., providing smallpox-laced blankets to indigenous people) that caused so much harm. As noted above, looking further ahead, we might also wish to consider how we will guide our actions when the scientific novelty wears off.

As there have been in the past, there are likely to be public interest groups attempting to ensure that space agencies are not only doing what is perceived to be environmentally prudent, but perhaps more broadly morally correct as well. Species preservations groups could have a new cause to champion, and it should be assumed that they will not hesitate to act if they have any reason to believe that the proper precautions are not being taken. Environmentalists opposing the use of nuclear power sources have been able to delay launches in the past. In this light, exploring the value

[23] See Cockell (2005) for an exploration of biocentrism and how it can be used to ascribe value to microorganisms.

of potential extraterrestrial life, particularly in the context of international deliberations (Rummel et al. 2012), and planning now to address the pre- and post-detection questions noted above, could help mitigate future challenges to sending humans to Mars.[24]

9.4.4 *Anthropocentric and Geocentric Bias?*

Ethics, understandably, has had a long history of human-centric and earth-centric biases. Like so many human pursuits, ethics is a practical tool. Ethics helps us get along with each other, and increasingly, with our environment—both of which contribute to human well-being. Applying that kind of ethical momentum to the possibilities of extraterrestrial life certainly seems like it would have serious limitations to say the least—as it appears to have had even on Earth when considering the moral status of terrestrial non-human animals which can inform our astrobiological deliberations (Cleland and Wilson 2013), including the possibilities for a broader "planetocentric ethic" that might ascribe intrinsic value to potentially life-bearing planets (Sullivan 2013).

Robert Haynes, Chris McKay, and Don MacNiven have been prompted by the consideration of extraterrestrial activities, including terraforming, to suggest the need for a "cosmocentric ethic." They conclude that existing ethical theories exclude the extraterrestrial environment because they are geocentric and cannot be applied to extraterrestrial environments, hence leaving a vacuum for a cosmocentric ethic. Martyn Fogg writes: "The concept of terraforming is inspiring enough to perhaps generate a formal effort toward extending environmental ethics to the cosmic stage." (Fogg 1995, 490).

Haynes says that anthropocentrism implies geocentrism because we know of no other sentient beings in the universe (Haynes 1990). Perhaps in the strictest sense, this is true for now because we only inhabit the earth, but presumably we can take our anthropocentrism anywhere we go. And we can still be anthropocentrists if we were to discover extraterrestrial intelligence. Haynes' claim does not seem to apply in a general sense. McKay notes that ecological ethics has been "inextricably intertwined" with life on Earth and so he comes to the same conclusion (McKay 1990). But this observation does not necessarily rule out the application of existing ethical theories to the question at hand. MacNiven, while offering no additional reasons, agrees with Haynes and McKay and further suggests that anthropocentrism, zoocentrism, and biocentrism would present no moral objection to activities such as terraforming. Nevertheless, some traditional ethical frameworks have been applied to the question of extraterrestrial life.

---

[24]For an analysis of social factors regarding Mars exploration, see Race (1995) Societal Issues as Mars Mission Impediments: Planetary Protection and Contamination Concerns.

9.4.5 Anthropocentrism and Kantian Ethics

Anthropocentrists would not have much reservation about displacing or possibly destroying indigenous extraterrestrial life (especially microbial life) if it was required for human exploration and settlement of an extraterrestrial environment—particularly under circumstances where great benefits to humanity were at stake (e.g., human survival). Anthropocentric ethical views make human needs and desires the priority, often at the expense of all else. As Robert Zubrin points out, there is the glaring problem that those who would say human settlement of Mars should not take priority over the continued existence of extraterrestrial microbes would also need to provide some explanation of why such a position would not apply to terrestrial microbes which we do not hesitate to kill regularly (Zubrin 1993). This is a reasonable challenge. However, it is also reasonable to suppose that extraterrestrial microbes should not be seen in quite the same light as terrestrial microbes. Zubrin himself acknowledges their unique value (Zubrin and Wagner 1996).

A response to Zubrin's challenge might be to point out that extraterrestrial microbes may not be harmful to our well-being, as are many terrestrial microbes. Perhaps extraterrestrial microbes should be assumed innocent until proven otherwise? Also, assuming Martian microbes are not of the same phylogenetic tree as life on earth, they would be unique in a way that terrestrial microbes are not. This significant uniqueness seems to imply some kind of unusual value (instrumental or intrinsic) that might not necessarily be attributed to terrestrial microbes. Cockell (2007) explores various applications of this thinking.

J. Baird Callicott's application of weak anthropocentrism suggests that the value of primitive extraterrestrial life can be justified by appealing to its transforming and ennobling effect on human nature. He writes: "I can think of nothing so positively transforming of human consciousness as the discovery, study, and conservation of life somewhere off the earth" (Callicott 1986a, 252). In addition, for Callicott, species possess a "truncated" version of the traditional definition of intrinsic value in that they have value "for" themselves, for their own sake, but not necessarily "in" themselves, independent of a valuing consciousness (Callicott 1986b).

The basis for Callicott's perspective on intrinsic value is what might be characterized as a "Humean-Darwinian" emotive-bioempathic view which suggests that emotionally based value identification with other living things results from natural selection. Furthermore, ethical relativism among humans can perhaps be avoided by appealing to Hume's "consensus of feeling" which standardizes or fixes the human psychological profile and many values that result. Although value may not be focused solely on humans in this view, humans are indeed the source of value in that they recognize intrinsic value of other living things.

But are values such as recognizing the intrinsic value of non-humans so standard or "fixed"? It appears not since there exists much controversy over the value of non-humans. Hence, there still appears to be an inherent subjectivity on an individual as well as a collective basis since the feelings of humans are what dictate the recognition of intrinsic value. This view, then, seems not to objectively justify intrinsic value or provide a way for measuring such value when difficult decisions have to be made.

A Kantian view, which could be considered to be a brand of anthropocentrism (in the sense that it could be limited to human rationality), or at least a kind of "ratio-centrism" (Smith 2009), appeals to the rational basis of ethics. Agents capable of rational thought are intrinsically valuable. Such a view would presumably ascribe no intrinsic value to primitive Martian microbes, although it might allow for their instrumental value to the extent that Martian microbes may be valuable to intelligent rational beings. A recognition of potential value to other rational beings may compel intelligent rational beings to be very careful about how they act in the wider universe, including potentially adversely affecting primitive forms of life if it is thought that such life-forms may be seen as value by other rational beings.

9.4.6 Rights

Carl Sagan wrote: "If there is life on Mars, I believe we should do nothing with Mars. Mars then belongs to the Martians, even if they are only microbes." His fuller context is helpful:

> A thorough reconnaissance will clearly occupy us for centuries. But there will be a time when Mars is all explored; a time after robot aircraft have mapped it from aloft, a time after rovers have combed the surface, a time after samples have been return safely to Earth, a time after human beings have walked the sands of Mars. What then? What shall we do with Mars?
>
> There are so many examples of human misuse of the Earth that even phrasing this question chills me. If there is life on Mars, I believe we should do nothing with Mars. Mars then belongs to the Martians, even if the Martians are only microbes. The existence of an independent biology on a nearby planet is a treasure beyond assessing, and the preservation of that life must, I think, supersede any other possible use of Mars. (Sagan 1980, 130)

Although Sagan does not explicitly invoke the notion of rights, his perspective can be thought as a rights-based ethics, which can be supported by ideas of intrinsic value (Cockell 2016; Smith 2016). But he also expresses concerns about human propensities to misuse environments and notes how that could compromise instrumental value or aesthetic value ("a treasure beyond assessing") that does not have to rely on intrinsic value or rights.

Haynes (1990) claims that Regan's (1976, 1983) "animal rights" view would ascribe rights to indigenous microbes. This is consistent with Chris McKay's view that is based on the intrinsic value of life principle and hence suggests that Martian microbes have a right to life—"to continue their existence even if their extinction would benefit the biota of Earth" (McKay 1990, 194). Presumably, this would involve minimal human presence on Mars if our presence was thought to pose a threat to indigenous life. The "interests" of organisms may be broadly seen as an important guideline and could include microbial organisms. Bernard Rollin draws a line between protozoa which exhibit behavior that might indicate consciousness (albeit broadly construed) and bacteria and plants for which there is no such evidence (Rollin 1981). Peters (2018) suggests a "Responsibility Ethics" toward all life because life is better than non-life since it can appreciate the "good" and as a result has intrinsic value.

Rights-based views face the challenge of explaining why primitive life should be intrinsically valuable and why microbes would have "rights" based on that value. Rights are problematic because even though rights are often intended to be "absolute", they are often seen as matters of degrees when difficult decisions have to be made. Degrees of rights, in the final analysis, ultimately seem no different than degrees of value. Indeed, J. Baird Callicott writes: "The assertion of 'species rights' upon analysis appears to be the modern way to express what philosophers call 'intrinsic value' on behalf of nonhuman species. Thus the question, 'Do nonhumans species have a right to exist?' transposes to the question, 'Do nonhuman species have intrinsic value?" (Callicott 1986b, 163). If one claims that non-human animals have intrinsic value and rights but that there are no degrees of intrinsic value or degrees of rights, how are we to reasonably assess situations that involve conflict of rights and/or interests between humans and other life-forms?

9.4.7 Hybrid Ethical Views

Gillett (1992) suggests a hybrid view combining anthropocentrism as applied to terrestrial activity combined with biocentrism toward worlds with indigenous life. Invoking a patchwork of theories to help deal with different domains and circumstances might be acceptable and perhaps even desirable especially when dealing with something as varied and complex as ethics—indeed, it has a certain commonsense appeal. However, we might consider Callicott's observation:

> But there is both a rational philosophical demand and a human psychological need for a self-consistent and all-embracing moral theory. We are neither good philosophers nor whole persons if for one purpose we adopt utilitarianism, another deontology, a third animal liberation, a fourth the land ethic, and so on. Such ethical eclecticism is not only rationally intolerable, it is morally suspect as it invites the suspicion of ad hoc rationalizations for merely expedient or self-serving actions. (Callicott 1986a, 251)

Another possible way to think about hybrid ethical views is to acknowledge the relevance of different narratives for different contexts. Cho (2009) explores the utility of Asian storytelling logic, noting that it recognizes important limitations of existing narratives and also that narratives are most useful when they are properly chosen for the context at hand. This kind of broad view is arguably consistent with what might be considered hybrid views in the sense that they could combine pieces of narratives to create the most useful narrative for novel contexts and/or that they might suggest numerous specific narratives for very specific contexts within a larger question or problem.

9.4.8 Formed Integrity and the Sanctity of Existence

Holmes Rolstonoffers a view which appeals to the "formed integrity" of a "projective universe." His view suggests that the universe creates objects of formed integrity—e.g., objects worthy of a proper name—which have intrinsic value and which should be respected (Rolston 1990). However, Robert Haynes points out that Rolston's view appears to conflict with modifying the Earth, even to the benefit of humans (Haynes 1990). Rolston's view would certainly call for the preservation of primitive extraterrestrial life.

Rolston's view is close to the conception of a cosmocentric ethic being explored here in that it seems to satisfy the first two criteria, and possibly the third and fourth, listed in the beginning of Sect. 9.4. In Rolston's view, justification of intrinsic value might come from the creative processes of the universe itself—that is, the creative process, and all that results from it, is intrinsic to the universe.[25] Rolston's view also attempts to address the problem of assessing or measuring value by suggesting that if a thing has formed integrity, or is worthy of a proper name, it should be respected. But how do we decide what has formed integrity so that it will be named? This is the value measurement problem in a different form. The fundamental challenge ultimately persists since personal subjective value judgments seem unavoidable in assessing what has formed integrity or is worthy of a proper name.

Don MacNiven has suggested that a central tenet of a cosmocentric ethic would be the "principle of the sanctity of existence," which he notes would make it difficult to justify the significant modification or destruction of indigenous life-forms (MacNiven 1995). In a minimal sense, the principle of the sanctity of existence seems to satisfy criteria one and two for the idea of a cosmocentric ethic noted previously because the universe, and all therein, *exists*. MacNiven additionally suggests appealing to a "selective concept of uniqueness" as we sometimes do in considering terrestrial matters such as preserving the Grand Canyon. In addition to not directly satisfying criterion three of the previous criteria of a cosmocentric ethic, we see still, even in light of the notion of uniqueness, the problem of "measuring" value—or more specifically, of weighing the value of human activity against other forms of value such as the preservation of an extraterrestrial life-form.

Somewhat related to the "sanctity of existence," others have applied Eastern views such as Buddhism and Hinduism to suggest that ideas such as (a) the principle of non-duality (emphasizing the "unity" of all existence and the de-emphasis of self), (b) "dependent co-arising," and (c) the common origin, interrelatedness, and interdependence of all beings for their existence, to suggest a certain degree of sensitivity and moral consideration to that which is seen as "totally other" (Irudayadason 2013).

[25]Similar to MacNiven's view of the principle of the sanctity of existence, Rolston's perspective has been referred to as "object-centered" by Richard Miller, in a 1996 independent study entitled, "The Greening of Mars: Ethics, Environment, and Society, Terraforming: An Ethical Perspective." University of Waterloo, Canada.

Here too we see broad suggestions seemingly implying some kind of intrinsic value and deep sensitivity for other life-forms, but without specific suggestions about how to assess what might be relative degrees of value to help guide practical behavioral actions and policies.

9.4.9 Complexity

As explored previously, complexity—possibly *increasing* complexity—may be one of many manifestations of the connection-action principle. This suggests we may want to consider the emerging science of complexity as a basis for a cosmocentric ethic, particularly when thinking about extraterrestrial life. A science of complexity could not only provide a scientific basis for value, but it could conceivably allow for relatively precise assessments, possibly even quantification, of degrees of value.

Smith (2014) ties rationality to a triad that also includes sociality and culture which ultimately gives rise to immense complexity, where complexity appears to be an overarching trend of biological evolution. Seen in totality, our particular biological evolution here on Earth can be interpreted to exhibit a trend from relative simplicity to increasing complexity. That complexity could at least be theoretically contingent on the particular twists and turns that evolution by natural selection took here on Earth, but as Smith suggests, it may also be indicative of universal trends for all life in the universe, and hence, could serve as a foundational ethic for all life and intelligence in the universe. Delahaye and Vidal (2018) go further and suggest that organized complexity might have intrinsic value and hence provide a "measurable" basis for a universal ethics.

The connection-action principle covered previously goes further in the sense that it goes beyond biology and organized complexity and suggests a conceptual theoretical underpinning for why the universe might trend in directions of organized complexity (among other possible trends that manifest connectedness via increasing degrees of relationships and actions), and hence may provide a basis for something like "intrinsic cosmological value" (e.g., intrinsic to cosmic evolution).

9.4.10 The Extraterrestrial Intelligence Connection

Acosmocentric ethic should provide a framework in which we can also explain and leverage any instrumental and aesthetic value of primitive extraterrestrial life-forms to intelligent beings. An appropriately conservative exploration approach seems to be a reasonable implication for a cosmocentric ethic, especially given the unknown aspects of interacting with extraterrestrial life. If something like the connection-action principle might provide guidance, then presumably other intelligent beings would be inclined to value life throughout the universe as a robust manifestation of the nature of the universe. A conservative exploration posture could help ensure that the value of such a discovery, including the possible value to other intelligent beings, would be part of the deliberations.

We might also note the possible importance of understanding value as it relates to extraterrestrial intelligence. An examination of value theory in this light may be useful by acting as a different lens through which we see values—as touched on in Chap. 3, "From Replication to Aspiration." Such an endeavor may help inform questions regarding interaction with extraterrestrial intelligence. Are there truly universal values, and would we need an understanding of them in order to interact constructively with extraterrestrial intelligence? Would other intelligent beings value microbial life on other planets for either instrumental or other reasons—some of which we might be able to imagine, but others not? Might other extraterrestrial intelligence beings value life throughout the universe because they are seen as cosmic creations that help increase cosmic diversity and so help manifest the nature of the universe?

9.4.11 Diversity and Peaceful Coexistence

Inlight of the above considerations, we can also consider the longer-term issue of whether to establish settlements and eventually fully populate Mars. Peaceful coexistence is one long-term option to consider as a thought experiment. Ironically, Richard Taylor's slogan, "Move over microbe!" might apply.[26] That is, extraterrestrial microbes might be displaced, as often happens on earth, but they need not be harmed or destroyed if we can figure out how to do that with high enough confidence—no trivial task to be sure. Can we coexist with Martian life? How important would it be to make sure we do not change its evolution too much, if at all? Would we combine into one ecosystem if we existed on the same planet together?

Callicott notes that coexistence may be feasible since we will not have to consume indigenous Martian life (Callicott 1986a). However, the needs or desires for human exploration and migration could cause many other conflicts of interest between humans and indigenous extraterrestrial life-forms.

If we think longer term, as unappealing as it may be to some, it could be that we humans, as the "dirty" biological beings we are now, will not be the ones settling Mars. Our much "cleaner" artificially intelligent descendants may be those who migrate to Mars and beyond (Morton 2018). If that comes to pass, coexistence might be more possible to the extent that impacts from our biology could be minimized or removed completely. Our robotic emissaries today may be the precursors of such a future—indeed, they arguably pose much less threat to any putative Martian life.

[26]Martyn Fogg (1995) notes a radio interview with Richard Taylor in *Terraforming: Engineering Planetary Environments*. SAE International, Warrendale, p. 494.

Assuming we were careful, Martian life might not be destroyed. It could, however, change via the forces of its new ecosystem. Or perhaps we will decide to preserve that life in a kind of isolated conservatory with the indigenous Martian environment intact, so that it will be allowed to evolve as it might have otherwise.[27] This may satisfy those who believe that primitive extraterrestrial life should evolve autonomously. The caveat of course would be to exercise extreme caution in our interaction with that environment, since most advocates of such a view would be highly skeptical of our ability to avoid causing harm or changes in that life-form, regardless of any good intentions.

For those who would suggest that Martian life has "rights," a peaceful coexistence compromise might not be satisfactory. Only a non-interference policy would be acceptable if there is a concern that our presence would compromise the existence of that life (Marshall 1993). However, we might also consider Chris McKay's view that the rights of Martian life "confer upon us the obligation to assist it in obtaining global diversity and stability" (McKay 1990, 194). This could arguably be consistent with cosmological reverence at least in the broad sense that cosmic creations (e.g., another kind of life-form) would be considered valuable as part of the evolution of the cosmos we revere. This kind of view might also be consistent with the connection-action principle and its implications for favoring diversity throughout the universe—assuming we were indeed capable of assisting microbial life on another planet. So if we interpret the connection-action principle to suggest that diversity is a manifestation of a form of cosmic value (perhaps a form of cosmic intrinsic value), then perhaps we can justify the preservation and "assistance" of other extraterrestrial life-forms based on a contribution to cosmic diversity?

> …silent… were the Martians — silent and alive! — preserved by the Humans against which their systems were otherwise unprepared… alive, after all the Human's devices had worked, alive from the care shown by those who had landed upon their world. And so, across the gulfs of space, minds that were to Humans as Humans were to the Martians, intellects vast and cool and sympathetic, regarded this Earth with admiration, and slowly and surely drew their plans to welcome us to the cosmic neighborhood—we who had evolved beyond our blind selfish genes—we who had chosen respect for the other. (Adapted from *War of the Worlds* by H.G. Wells)[28]

9.5 Summary of Chapter 9: Meaning and Ethics

In Part 2, we explored three cosmological theories of value: cosmological reverence, cosmocultural evolution, and the connection-action principle. These three "cosmological worldviews" ascribe various forms and degrees of value to the universe and to life and intelligence, informed by the context of cosmic evolution. We briefly

---

[27] Robert Zubrin opens the door for such a compromise when he suggests that the polar regions could be available for indigenous life to populate (Zubrin 1993).

[28] Adapted from *War of the Worlds* by H. G. Wells. Originally in Lupisella and Logsdon (1997).

touched on how these worldviews ascribe value to a morally creative cosmos, one in which we can emphasize respect for relationships, diversity, and complexity.

Cosmological reverence suggests "unidirectional" value in the sense that the universe is valuable for beings like ourselves who can look back on our cosmic evolutionary history and choose to value that cosmological relationship. This may then engender some kinds of ethical views for the universe as a whole, but creatures like us would not have value for the universe as a whole.

Cosmocultural evolution, however, suggests that not only is the universe valuable to beings such as ourselves which have emerged from cosmic evolution, but also that we have value and significance for the universe. Intentional, contemplative, and cultural valuing agents can serve as a means by which the universe has bootstrapped itself into the realms of value, meaning, purpose, and ethics. This increases our ethical commitments to the universe in the sense that we have greater responsibility to act as arbiters of value, meaning, and purpose for the cosmos—to act as a kind of cerebral cortex for the universe, including possibly working to ensure the eternal existence of this universe and perhaps the creation and perpetuation of other universes as well.

The *connection-action principle* takes us quite a bit further and makes the speculative claim that the dynamic and creative nature of the universe is grounded in the universe's property of connectedness which is manifested as relations and actions—perhaps in ever-increasing degrees. This suggests a kind of intrinsic cosmological value for relationality and action, and "derivitaves" such as diversity, freedom, complexity, and any other phenomenon that helps realize the universe's nature of connectedness through relations and action.

We applied these broad cosmological theories of value to explore cosmic ethics, or more strongly, "cosmocentric" ethics, specifically applied to the question of putative life on Mars. We leveraged the cosmic values of the connection-action principle such as relationship, diversity, and complexity to suggest that even "primitive" Martian life is worthy of our respect and possible ethical commitment. This might take the form of something like "peaceful coexistence", which might even be more justified if Martian life is an independent origin of life since it would arguably significantly enhance the diversity of the universe.

We briefly revisited the potential applicability of the "psychology of philosophy," heightening our sensitivity to the potentially aesthetic appeal (possibly misleading in some cases) of notions of connectedness and diversity, which may have evolutionary roots. That psychological sensitivity, along with more detailed practical ethical considerations, led us to raise questions about the pragmatic utility of the connection-action principle and whether assessing degrees of value in its context would be credible. Nevertheless, we briefly considered at least theoretical paths forward for such a possibility.

References

Aerts, D., Van Belle, H., & Van der Veken, J. (1994). *Worldviews: From fragmentation to integration*. Brussels: VUBPRESS.

Aerts, D., Van Hubert, B., & Van der Veken, J. (1999). *World views and the problem of synthesis*. Dordrecht: Kluwer Academic.

Almar, I., & Schuch, P. (2007). The san marino scale: A new analytical tool for assessing transmission risk. *Academy Transactions Note. Acta Astronautica, 60*, 57–59.

Babbage, D. R., & Ronan, K. R. (2000). Philosophical worldview and personality factors: Studying the world in our own image. *Personality and Individual Differences, 28*(2), 405–420.

Barlow, C. (1997). *Green space, green time: The way of science*. New York: Springer.

Bobskill, M., & Lupisella, M. (2014). Human mars surface science operations. In: *SpaceOps 2014 Conference Paper AIAA-2014-1620*. Pasadena, CA: American Institute for Aeronautic and Astronuatics.

Bulbulia, J., Sosis, R., Harris, E., Genet, R., Genet, C., & Wyman, K. (Eds.). (2008). *The evolution of religion: Studies, theories, and critiques*. Santa Margarita (CA): Collins Foundation.

Burkhart, B. Y. (2004). What Coyote and Thales can teach us: An outline of American Indian epistemology. In A. Waters (Ed.), *American Indian thought*. Mauldin: Blackwell.

Callicott, J. B. (1986a). Moral considerability and extraterrestrial Life. In E. C. Hargrove (Ed.), *Beyond spaceship earth: Environmental ethics and the solar system*. San Francisco: Sierra Club Books.

Callicott, J. B. (1986b). On the intrinsic value of Nonhuman species. In B. Norton (Ed.), *The preservation of species* (pp. 138–172). Princeton University: Princeton.

Cho, F. (2009). Comparing stores about the origin, extent, and future of life: An Asian religious perspective. In C. Bertka (Ed.), *Exploring the origin, extent, and future of life: Philosophical, ethical and theological perspectives* (pp. 303–320). Cambridge: Cambridge University.

Cleland, C. E. (2019). *The Quest for a Universal Theory of Life: Searching for Life As We Don't Know It*. United Kingdom: Cambridge University Press.

Cleland, C. R., & Wilson, E. M. (2013). Lessons from earth: Toward an ethics of astrobiology. In C. Impey, A. H. Spitz, & W. Stoeger (Eds.), *Encountering life in the universe: Ethical foundations and issues and social implications* (pp. 17–55). Tucson: University of Arizona.

Cockell, C. S. (2005). The value of microorganisms. *Environ Ethics, 27*, 375–390.

Cockell, C. S. (2007). Originism: Ethics and extraterrestrial life. *JBIS, 60*, 147–153.

Cockell, C. S. (2016). The ethical status of microbial life on earth and elsewhere: In defence of intrinsic value. In J. Schwartz & T. Milligan (Eds.), *The ethics of space exploration: Space and society*. Cham: Springer.

Committee on Space Research (COSPAR) Planetary Protection Policy, International Council for Science. (2015). *Space research today*, COSPAR's information bulletin, Number 193, August 2015). https://cosparhq.cnes.fr/sites/default/files/ppp_article_linked_to_ppp_webpage.pdf.

Cypser, D. A. (1993). International law and policy of extraterrestrial planetary protection. *Jurimetrics, 33*, 315.

Davies, P. C., Benner, S. A., Cleland, C. E., Lineweaver, C. H., McKay, C. P., & Wolfe-Simon, F. (2009) Signatures of a Shadow Biosphere. *Astrobiology, 9*(2), 241–249.

Delahaye, J. P., & Vidal, C. (2018). Universal ethics: Organized complexity as an intrinsic value. In G. Y. Georgiev, J. M. Smart, C. L. Martinez, & M. E. Price (Eds.), *Evolution, development and complexity: Multiscale evolutionary models of complex adaptive systems*. Berlin: Springer.

Dick, S. J. (1995). Consequences of success in SETI: Lessons from the history of science. In S. Shostak (Ed.), *G* (pp. 521–532)., Progress in the search for extraterrestrial life San Francisco: Astronomical Society of the Pacific.

Dick, S. J. (Ed.). (2015). *The impact of discovering life beyond earth*. Cambridge: Cambridge University.

Dick, S. J. (Ed.). (2018). *Astrobiology, discovery, and societal impact*. Cambridge: Cambridge University.

Ehrenfreund, P. et al. (2013). *COSPAR workshop on developing a responsible environmental regime for celestial bodies*. Committee on Space Research of the International Council for Science and The Space Policy Institute, George Washington University. https://www.gwu.edu/~spi/assets/docs/PEX_WorkshopReport_ES_March14_Web%20(1).pdf.

Feibleman, J. K. (1967). Normative cosmic ethics. In: *Moral strategy*. Dordrecht: Springer.

Fogg, M. (1995). *Terraforming: Engineering planetary environments*. Warrendale: SAE International.

Gershenson, C., Aerts, D., & Edmonds, B. (Eds.). (2007). *Worldviews, science and us: Philosophy and complexity*. Singapore: World Scientific Publishing.

Gillett, S. (1992). The ethics of terraforming. *Amazing* 72–74.

Glavin, D. P., Dworkin, J. P., Lupisella, M. L., Kminek, G., & Rummel, J. D. (2004). Biological contamination studies of lunar landing sites: Implications for future planetary protection and life detection on the Moon and Mars. *Int J Astrobiol, 3*(3), 265–271.

Green, B. P. (2020). Convergences in the Ethics of Space Exploration. In *Social and Conceptual Issues in Astrobiology*, Kelly C. Smith & Carlos Mariscal, (eds.), New York: Oxford University Press.

Haidt, J. (2012). *The righteous mind*. New York: Random.

Haynes, R. (1990). Ecopoiesis: Playing god on Mars. In D. MacNiven (Ed.), *Moral expertise*. London: Routledge.

Haynes, R. H., & McKay, C. P. (1990). Should we implant life on mars? In: MacNiven's, *Creative Morality*, p. 144. Scientific American.

Hedlund-de Witt, A., de Boer, J., & Boersema, J. J. (2014). Exploring worldviews and their relationships to sustainable lifestyles: Towards a new conceptual and methodological approach. *Ecological Economics, 84,* 74–83.

Henriques, G. (2011). *A new unified theory of psychology*. Berlin: Springer.

Impey, C., Spitz, A., & Stoeger, W. (Eds.). (2013). *Encountering life in the universe: Ethical foundations and social implications of astrobiology*. Tucson: The University of Arizona Press.

Irudayadason, N. (2013). The wonder called cosmic oneness: Toward astroethics from Hindu and Buddhist wisdom and worldviews. In C. Impey, A. Spitz, & W. Stoeger (Eds.), *Encountering life in the universe: Ethical foundations and social implications of astrobiology* (pp. 94–119). Tucson: University of Arizona Press.

Johnson, K. A., Hill, E. D., & Cohen, A. B. (2011). Integrating the study of culture and religion: Toward a psychology of worldview. *Social and Personality Psychology Compass, 5,* 137–152.

Jonas, H. (1984). *The imperative of responsibility*. Chicago: University of Chicago.

Kminek G. et al. (2016). Report of the COSPAR Workshop on Refining Planetary Protection Requirements for Human Missions. https://sma.nasa.gov/sma-disciplines/planetary-protection

Koltko-Rivera, M. E. (2004). The psychology of worldviews. *Review of General Psychology, 8*(1), 3–58.

Lupisella, M. (1997). The rights of Martians? *Space Policy, 13*(2), 89.

Lupisella, M. (1999). Ensuring the scientific integrity of possible Martian life. In: *Paper IAA-99-IAA.13.1.08 Presented at the International Astronautical Federation Congress*. Amsterdam: American Institute of Aeronautics and Astronautics.

Lupisella, M. (2004). Using artificial life to assess the typicality of terrestrial life. *Advances in Space Research, 33,* 1318–1324.

Lupisella, M. (2019). Is the universe enough? Can It suffice as a basis for worldviews? *Journal of Big History, 3*(3), 123–140.

Lupisella, M., & Logsdon, J. (1997). Do we need a cosmocentric ethic? In: *Paper IAA-97-IAA.9.2.09 Presented at the International Astronautical Federation Congress*. Turin: American Institute of Aeronautics and Astronautics.

Lupisella, M., & Race, M. (2017). Low-latency teleoperations, planetary protection and astrobiology. *International Journal of Astrobiology, 17,* 1–8. https://doi.org/10.1017/s1473550417000374.

Lupisella, M. L., Wright, M. R., Bleacher, J. E., Young, K. E., Gernhardt, M. L., Chappell, S. P., & Beaton, K. H. (2017). Low-latency teleoperations and telepresence for the evolvable Mars campaign. 2017 IEEE Aerospace Conference, 1–17.

MacNiven, D. (1995). Environmental ethics and planetary engineering. *Journal of the British Interplanetary Society, 48,* 442–443.

Marshall, A. (1993). Ethics and the extraterrestrial environment. *Journal of Applied Philosophy, 10*(2), 227.

McKay, C. (1990). Does Mars have rights? In D. MacNiven (Ed.), *Moral expertise*. London: Routledge.

McKay, C., & Davis, W. (1989). Planetary protection issues in advance of human exploration of Mars. *Advanced Space Research, 9*(6), 197.

Miller, R. (1996). *The greening of Mars: Ethics, environment, and society, terraforming: An ethical perspective, independent study entitled*. Canada: University of Waterloo.

Morton, A. (2018). *Should we colonize other planets?*. Cambridge: Polity Press.

Murphy, N., & Ellis, G. (1996). *On the moral nature of the universe; theology, cosmology, and ethics*. Minneapolis: Fortress Press.

Nilsson, A. (2013). *The psychology of worldviews: Toward a non-reductive science of personality* (Doctoral dissertation, Lund University, Sweden). Retrieved 2018 11 14 from http://lup.lub.lu.se/record/3738772.

Peters, T. (2018). Does extraterrestrial life have intrinsic value? An exploration in responsibility ethics. *International Journal of Astrobiology, 17*(2), 1–7.

NASA Policy on Planetary Protection Requirements for Human Extraterrestrial Missions. https://planetaryprotection.arc.nasa.gov/documents.

Race, M. S. (1995). Societal issues as Mars missions impediments: Planetary protection and contamination concerns. *Advanced Space Research, 15,* 285.

Race, M. S. (2013). Space exploration and searches for extraterrestrial life: Decision making and societal issues. In C. Impey, A. Spitz, & W. Stoeger (Eds.), *Encountering life in the universe: Ethical foundations and social implications for astrobiology*. Tucson: The University of Arizona Press.

Race, M. et al. (2015). NASA Workshop on Planetary Protection Knowledge Gaps for Human Extraterrestrial Missions. https://sma.nasa.gov/sma-disciplines/planetary-protection.

Race, M. et al. (2018). 2nd COSPAR WORKSHOP on Refining Planetary Protection Requirements for Human Missions. https://sma.nasa.gov/sma-disciplines/planetary-protection.

Race, M. et al. (2019). 3rd COSPAR Meeting on Refining Planetary Protection Requirements for Human Missions. https://sma.nasa.gov/sma-disciplines/planetary-protection.

Race, M. S., & Randolph, R. (2002). The need for operating guidelines and a decision-making framework applicable to the discovery of non-intelligent extraterrestrial life. *Advances in Space Research, 30,* 1583–1591.

Regan, T. (1976). The case for animal rights. In T. Regan & P. Singer (Eds.), *Animal rights and human obligations*. Englewood Cliffs: Prentice Hall.

Regan, T. (1983). *The case for animal rights*. Berkeley: University of California Press.

Rescher, N. (1985). *The strife of systems: An essay on the grounds and implications of philosophical diversity*. Pittsburg: University of Pittsburg Press.

Rollin, B. (1981). *Animal rights and human morality*. Buffalo: Prometheus Books.

Rolston, H. (1990). The preservation of natural value in the solar system. In E. C. Hargrove (Ed.), *Beyond spaceship earth: Environmental ethics and the solar system*. San Francisco: Sierra Club Books.

Rucker, M. (2013). *Drilling system study, mars design reference architecture 5.0.* NASA/JSC/Human Exploration Development Support, Document JSC-66635. Release Date: September 30, 2013.

Rummel, J. D., Race, M. S., & Horneck, G. (2012). Ethical considerations for planetary protection in space exploration: A workshop. *Astrobiology, 12*(11), 1017–1023.

Sagan, C. (1980). *Cosmos*. New York: Random House.

Saucier, G. (2013). Isms dimensions: Toward a more comprehensive and integrative model of belief-system components. *Journal of Personality and Social Psychology, 104*(5), 921–939.

Saucier, G., Kenner, J., Iurino, K., Malham, P. B., Chen, Z., Thalmayer, A. G., et al. (2015). Cross-cultural differences in a global "survey of world views". *Journal of Cross-Cultural Psychology, 46*(1), 53–70.

Schwartz, J. S. J., & Milligan, T. (Eds.). (2016). *The ethics of space exploration, space and policy series*. Berlin: Springer.

SETI Permanent Committee. (2010). *Declaration of principles concerning the conduct of the search for extraterrestrial intelligence*. International Academy of Astronautics. http://iaaweb.org/iaa/Scientific%20Activity/setideclaration2.pdf.

Shermer, M. (2002). *Why people believe weird things*. New York: Henry Holt and Company.

Smith, K. C. (2009). The trouble with intrinsic value: A primer for astrobiology. In C. Bertka (Ed.), *Exploring the origin, extent, and future of life: Philosophical, ethical and theological perspectives* (pp. 261–280). Cambridge: Cambridge University Press.

Smith, K. (2014). Manifest complexity: A foundational ethic for astrobiology? *Space Policy, 30*, 209–221.

Smith, K. C. (2016). The curious case of the Martian microbes: Mariomania, intrinsic value and the prime directive. In J. Schwartz & T. Milligan (Eds.), *The ethics of space exploration, space and society*. Berlin: Springer.

Smith, K. C. & Mariscal, C., (Eds.) (2020). Social and Conceptual Issues in Astrobiology. New York: Oxford University Press.

Sontag, S. (2007). *At the Same Time: Essays and Speeches*. Eds. Paolo Dilonardo and Anne Jump. New York: Farrar, Straus, Giroux. Original Essay, "The Truth of Fiction Evokes Our Common Humanity", given by Susan Sontag April 7, 2004 at the Los Angeles Public Library upon receiving the library's Literary Award. Published in the Los Angeles Times, December 29, 2004.

Sullivan, W. T. (2013). Planetocentric ethics: Principles for exploring a solar system that may contain extraterrestrial microbial life. In C. Impey, A. H. Spitz, & W. Stoeger (Eds.), *Encountering life in the universe: Ethical foundations and issues and social implications* (pp. 167–177). Tucson: University of Arizona Press.

Vakoch, D. A. (Ed.). (2011). *Communication with extraterrestrial intelligence*. Albany: State University of New York Press.

Vidal, C. (2012). Metaphilosophical criteria for worldview comparison. *Metaphilosophy, 43*(3), 306–347.

Vidal, C. (2014). *The beginning and the end: The meaning of life in a cosmological perspective*. Berlin: Springer.

Wilson, D. S. (2002). *Darwin's cathedral: Evolution, religion, and the nature of society*. Chicago: The University of Chicago Press.

Wilson, E., & Cleland, C. (2015). The moral subject of astrobiology: Guideposts for exploring our ethical and political responsibilities towards extraterrestrial life. In S. J. Dick (Ed.), *The impact of discovering life beyond earth* (pp. 207–221). Cambridge: Cambridge University Press.

Zubrin, R. (1993). The terraforming debate. *Mars Underground News, 3*, 3–4.

Zubrin, R., & Wagner, R. (1996). *The case for Mars* (pp. 134–135). New York: The Free Press.

Chapter 10
Synthesis and Summary

10.1 Brief Summary

The Preface began with an acknowledgment of sensitivities we should have when engaged in worldview-building, ranging from misguided psychological predispositions to problematic practical implications when worldviews are misused. Chapter 1, Overview and Road Map, provided some motivation for this book's exploration along with a graphical overview or partial "road map" of the book showing how the various parts fit and relate to each other. A slightly expanded version of that diagram is in this summary chapter as Fig. 10.1, capturing a few other details that were explored previously. We acknowledged the many difficulties in contemplating long-term cosmic visions, but nevertheless suggested it is worthwhile, particularly given what we continue to learn about the universe and our origins, and given that scientifically informed philosophical speculation can have surprising relevance and utility. Exploring alternative worldviews can be helpful personally and collectively and can help inform not only detailed philosophical considerations (such as ethics) but potentially scientific pursuits as well. Science and philosophy can often be more closely tied than it may appear, and this book explored some of those relationships.

In the first half of the book, Foundational Context (Chaps. 2–5), we explored a broad foundational context involving scientific minimalism and relationalism to help inform the second part of the book on Cosmological Worldviews and Implications (Chaps. 6–9) that developed three cosmological theories of value of arguably increasing consequence: *Cosmological Reverence*, *Cosmocultural Evolution*, and *the Connection-Action Principle*. These views provide a broad cosmic context for various forms of meaning for life and intelligence in the universe, which can also help inform "cosmic ethics."

Chapter 2 explored some past and present perspectives that relate to life and intelligence in a cosmic context. Chapter 3 then briefly covered key aspects of life (e.g. natural selection) and intelligence that relate to exploring what life and intelligence might mean in a cosmic context, including some brief reflection on culture, value theory, and the potentially uniquely compelling role for *normative aspiration* of

M. Lupisella, *Cosmological Theories of Value*, Space and Society,
https://doi.org/10.1007/978-3-030-25339-4_10

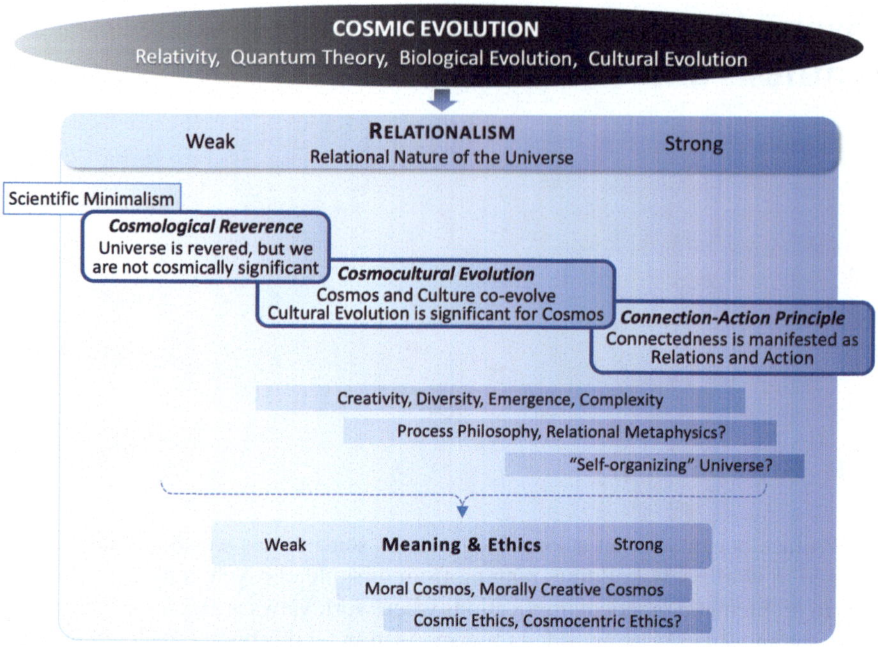

Fig. 10.1 Summary diagram (details added to Fig. 1.1)

intelligent beings in cosmic evolution. Chapter 4 then shifted to a more in-depth examination of the modern sciences of relativity, quantum theory, and cosmology to (1) help inform the development of the subsequent relationalist framework and three cosmological worldviews, and (2) to also suggest a form of "scientific minimalism" that suggests a minimalist scientific worldview can reserve judgment and/or reject many broader philosophical or interpretive judgments that do not need to go beyond scientifically operational knowledge.

A relational framework was developed in Chap. 5, drawing from details of modern physics (e.g. relativity and quantum theory) and cosmic evolution (including biological and cultural evolution), and relational philosophy more broadly. This relationalist framework, or "relationalism," emphasized the importance of relationality in general and relations in particular as a philosophical framework that can help bridge across science and philosophy. Relationalism then provided a broader context and pluralistic philosophical framework for the development of three cosmological theories of value: *Cosmological Reverence* in Chap. 6, *Cosmocultural Evolution* in Chap. 7, and *The Connection-Action Principle* in Chap. 8. Chapter 9 further explored some implications for meaning and ethics.

Cosmological reverence ascribes a significant level of value to the universe based on the scientific view that life and intelligence are fundamentally related to the physics of the universe and as a direct product of cosmic evolution. *Cosmocultural evolution* goes further and suggests that culture co-evolves with the universe in an open-ended and potentially meaningful and purposeful manner, influenced heavily, if not eventually completely, by the values and aspirations of valuing agents such as cultural beings that may ultimately have significance for, and affect the evolution of, the universe. Finally, within the context of scientifically and philosophically informed relationalism, we explored the *connection-action principle* that offers what may be a fundamental conceptual foundation for why the universe is dynamic and creative, and perhaps increasingly so, namely that *the universal property of connectedness is manifested as relations and actions*.

Based on the three cosmological theories of value, we further explored implications for value, meaning, and ethics and applied the theories to a few questions regarding interaction with extraterrestrial life and intelligence—with varying degrees of implications for moral and ethical commitments based on weaker and stronger forms of value and the roles of life and intelligence for the universe and its evolution. Cosmological reverence is the least prescriptive basis for cosmic value, in part because valuing agents such as ourselves do not have any notable or compelling significance for the universe. Nevertheless cosmological reverence supports basic forms of respect for cosmic creations, including extraterrestrial life. Cosmocultrual evolution however suggests that intelligent cultural beings such as ourselves do have significance for the universe, including the possibility that cultural beings are a means by which the universe is bootstrapped into realms of value, meaning, and purpose. Such beings are the co-creators and arbiters of value and meaning via cultural evolution which may ultimately be the most significant "force" in cosmic evolution. This role confers upon such beings a unique and compelling sense of responsibility to carefully contemplate our worldviews and actions, potentially on behalf of the universe itself.

The connection-action principle moves us beyond the subjective values of intelligent valuing agents and suggests a conceptual underpinning for how value might be grounded in the nature of the universe as the realization of its connectedness through relations and actions in a myriad of ways ranging from diversity to complexity. In such a view, many cosmic creations can be valued for cosmic reasons that do not rely only on the created values of intelligent beings. Such beings must then still nevertheless carefully explore the complex, uncertain, and perhaps endless "open-ended" possibilities and implications for how we see and realize potentially "universal" values in the forms of relationality and action.

Suggestions of, or worse, blind attachments to "cosmic" values should certainly raise flags. There is a dangerous history of developing dogmatic worldviews that claim to transcend human agency. Such transcendence is often a double-edged sword. Many forms and consequences of world religions and other "cult-like" worldviews are an obvious reminder of the potential pitfalls. Perceived "universal" values can bind but also blind.[1] Such perceptions may sometimes be helpful but can also be

[1] Related, Jonathan Haidt notes that morality can bind, but also blind, acting as a force to motivate groups against each other (Haidt 2012). A "universal" set of values, while intended to transcend that kind of parochial group function, might still nevertheless benefit from a sensitivity to a potentially blinding affect for those claiming universal values.

dangerously misleading. In some cases, we can know the difference. In others cases, we can not. But a truly systemic view, a cosmic view, does not have to set us up for a false tension between here and there, inward and outward, local and far, now and later. A truly systemic view, for example, a relationalist framework of the kind explored in this book, can reduce those tensions. Considerations of such a totality, of a more holistic view, however complex and daunting, seems justified as an intellectual pursuit.

10.2 Summary of Part I: Foundational Context

10.2.1 Summary of Chapter 2: Precedent

In Chap. 2, "Precedent," we briefly explored some speculative scientific and philosophical worldviews that are relevant for the kinds of worldviews explored in this book. We touched primarily on ideas that draw from modern science and cosmology and explored specific views regarding their potential relevance for life and intelligence in the universe. Views explored were in categories such as: (1) cosmological eschatology worldviews that emphasize the very far future of cosmic evolution and suggests that intelligence will evolve eventually to the point where it in some sense merges with the universe and possibly create a kind of eternal cosmic mind of sorts, (2) cosmic reverence worldviews that emphasize a deep reverence for the cosmos and its evolution as a source of our origins and being, (3) evolutionary developmental views that emphasize the evolutionary developmental nature of the universe and how it may give rise to other universes through the evolution of intelligence, (4) participatory universe views that leverage extreme interpretations of quantum theory to suggest that observation and consciousness gives rise to reality, including possibly the laws of physics themselves, and (5) science fiction, e.g., Olaf Stapledon's *Star Maker* that explores the idea of a "creator" imperfectly creating many universes and beings in an attempt to constantly improve on previous creations.

10.2.2 Summary of Chapter 3: From Replication to Aspiration

In Chap. 3, "From Replication to Aspiration," we briefly touched on key scientific ideas such as natural selection and how it has led to intelligence and cultural evolution—and how those developments ultimately gave rise to the normative aspiration of contemplative cultural beings like ourselves. We explored some fundamental areas of value theory such as the fact–value interplay and how the values of cultural agents could ultimately become a major factor in cosmic evolution. We also touched on the distinction between instrumental value and intrinsic value, noting the relational nature of instrumental value and the difficulties of justifying intrinsic value. Finally, we broached the subject of the psychology of philosophy, sensitizing ourselves to how psychological predispositions (likely caused by biological evolutionary selective pressures) can motivate and possibly mislead us in the pursuit of finding patterns, meaning, purpose, and in building sweeping "theories of everything."

10.2.3 Summary of Chapter 4: Scientific Minimalism

In Chap. 4, "Scientific Minimalism," we examined a limited set of philosophical implications based on the present state of modern science—with a heavy emphasis on quantum theory. It was suggested that scientific minimalism would generally refrain from making strong philosophical or prescriptive claims regarding cosmic meaning and purpose. However, stronger versions of scientific minimalism could suggest that science can and perhaps should help inform what values might be best for human beings and how we might go about pursuing those values—at least partially, but not exclusively, informed by evolutionary psychology. Scientific minimalism, again with an emphasis on evolutionary psychology, might also suggest caution regarding the appeal of certain philosophical ideas and values and our pursuits to find patterns, meaning, and purpose.

Scientific minimalism suggests that quantum theory as we understand it today is almost certainly incomplete and that more unified theories should be pursued. However, scientific minimalism can also suggest that too much speculation or concern about broader conceptual or intuitive implications and interpretations of quantum mechanics might not be as important as some think. Epistemological prudence and functionalism are important and can be leveraged in the pursuit of increasing verisimilitude despite acknowledged challenges (and perhaps fundamental limitations) of knowing the complete ontological truth about the world. Our foray into the Copenhagen interpretation of quantum mechanics suggests that scientific minimalism is at least agnostic about its interpretive value if not dismissive of its broader philosophical claims.

Questioning the Copenhagen interpretation also then calls into question the stronger cosmological claims that are sometimes derived from it. For example, suggestions that intelligent observers play a critical role in bringing the universe into being, and also potentially bringing physical laws themselves into being, would be considered suspect. While scientific minimalism would recognize a critical causal relationship of life and human intelligence to cosmic evolution as products of cosmic evolution, it would generally not imbue that relationship with any particular *value* per se—other than to recognize any basic value that may be associated with the physical causes of our existence. That is, our emergence from cosmic evolution is notable and important for understanding the world and possibly important in recognizing a relationship between ourselves and the universe, but that cosmic linkage is not necessarily important for a detailed exploration of human values.

The scientific minimalist perspectives of value, quantum theory, and cosmic evolution do however suggest some utility for thinking in terms of relationships and relationality more generally as a kind of broad framework or "metaview" of sorts. Indeed, we saw that science can relate to value by informing us about facts of the world and also by noting how at least instrumental value is essentially relational in the sense that value is usually realized in relation to something else. We also saw that the relational interpretation of quantum mechanics appears to have substantial merit. Scientific minimalism is also arguably consistent with the importance of *process*, at least in the form of physical dynamics that unfold in the universe over time.

10.2.4 Summary of Chapter 5: Relational Framework

In Chap. 5, "A Ralational Framework," we explored specific philosophical and scientific ideas that inform a relational framework. The relational nature of process philosophy was highlighted and we explored the idea that relativity and quantum theory were "relational revolutions". We briefly noted relational aspects of biology, psychology, and cultural evolution and suggested that cosmic evolution can be seen as a highly complex web of dynamics and relations. As Figs. 1.1 and 10.1 illustrate, this relationalist context serves as a broad framework within which to think about stronger versions of scientific minimalism and to help inform more philosophical worldviews such as *cosmological reverence*, *cosmocultural evolution,* and the *connection-action principle* developed in Part II.

10.3 Summary of Part II: Cosmological Worldviews

10.3.1 Summary of Chapter 6: Cosmological Reverence

One way to think about the relationships between the universe and intelligence is to explore how each is important for the other. Unidirectional or "one-way" relationships suggest that the universe is important for intelligence and culture but not the reverse. As shown in Fig. 6.1, this could be consistent with many views such as a biotolerant and biofriendly universe, as well as a "weak bootstrapped universe" worldview that suggests the universe has bootstrapped itself into the realm of value via cultural beings, but without any additional broader significance beyond that for the universe at large.

In Chap. 6, we drew from contemporary science and philosophy of previous chapters in Part I to briefly develop *cosmological reverence*—the first and arguably "weakest" of three cosmological worldviews that fit into a broader relational framework or relational philosophy (as shown in Fig. 1.1). Cosmological reverence is a weaker "unidirectional" relational view in the sense that while it emphasizes our fairly well-understood scientific relationships to the universe (namely that life and intelligence are direct products of cosmic evolution) and ascribes a certain value to those relationships, cosmological reverence nevertheless does not go further and suggest that intelligent beings have any significance for the universe as a whole.

Nevertheless, cosmological reverence can provide a basis for valuing the universe and cosmic creations associated with the dynamic relational process of cosmic evolution. We explored normative aspiration in this broader context, with an emphasis on both natural selection and cosmic evolution and what it might mean philosophically and more pragmatically—including brief treatments of implications for potentially very powerful and long-lived beings that may see their normative aspirations as ultimate drivers of the realities they create and the wisdom they seek.

10.3.2 Summary of Chapter 7: Cosmocultural Evolution

In Chap. 7, "Cosmocultural Evolution," we explored "bidirectional" relationships between intelligent cultures and the cosmos suggesting not only that the universe is important for culture but that intelligence and culture are important for the universe. This could include worldviews that can be characterized as a "strong bootstrapped universe," teleological, pantheistic, and theistic—all of which could be consistent with cosmocultural evolution or the "cosmocultural principle", which suggests that cultural evolution is significant enough for the cosmos that it implies an important coevolution of cosmos and culture that should be considered holistically as a new kind of integrated evolution.

Cosmocultural evolution emphasizes that a kind of "emergent property" of value has emerged in the universe in the minds of beings with interests that gives rise to that value, along with the more much broader and complex phenomenon of culture that collectively operationalizes such value and adds a significant, and arguably qualitatively different, kind of evolution to the cosmic landscape. *Bootstrapped cosmocultural evolution* suggests that the universe has "bootstrapped" itself into the realm of value via physical processes that created replicators leading eventually to intelligence and culture—but none of those phenomena were necessarily inherent in the universe (e.g., they were not a "cosmic imperative"). Nevertheless, this view suggests that value, meaning, and purpose now have a notable kind of cosmic significance and practical cultural relevance. In stronger versions of bootstrapped cosmocultural evolution, there is an unlimited potential for cultural beings to eventually influence and possibly significantly transform the whole universe. This emergence of a new kind of cosmic property, *value*, along with the cultural evolution that instantiates that value and creates derivatives of that value such as meaning and purpose (and other potentially endless forms of value), has given rise to a qualitatively different kind of cosmic evolution that may be completely open-ended and have unlimited potential.

This potentially unlimited power of cultural evolution presents profound responsibilities and challenges to sufficiently intelligent beings that may become increasingly relevant on ever-widening scales as cultural evolution begins to become a more cosmically relevant phenomenon. The forces of morality and creativity may give rise to a morally creative cosmos, a universe that goes beyond intelligence and technology and is driven more by the caring capacity of valuing agents and ultimately by an emerging pervasive cosmic force of moral creativity—something to which all cultural beings might ultimately aspire.

10.3.3 Summary of Chapter 8: The Connection-Action Principle

In Chap. 8, "The Connection-Action Principle", we moved quite a bit further along the relational spectrum in Figs. 1.1 and 10.1 and introduced the *connection-action principle* (CAP) which suggests the universe's property of connectedness is manifested by relations and actions, as captured graphically by Fig. 10.2, (a slightly simpler version of Fig. 8.1). Stronger interpretations of the connection-action principle suggest that relations and actions are *necessary* manifestations of connectedness, perhaps in *ever-increasing* degrees.

CAP is consistent with, and may help provide a conceptual foundation for some of the relational interpretations of relativity and quantum theory as well as providing a possible explanation for the ongoing creativity of cosmic evolution. CAP is also consistent with, and may offer a foundation for a number of philosophical ideas such as process philosophy, the principle of plenitude, and "organic unity," intrinsic value. Energy, time, and physical "laws" may be building blocks for instantiating connectedness in myriad forms or relations and actions. While "weaker" versions of CAP are largely agnostic about the distinction between interaction and *intra*-action, it was suggested that *intra*-action may be a better interpretation for CAP, suggesting a fundamental relational metaphysics more generally.

A number of general and admittedly speculative "predictions" of CAP were suggested, such as: (a) life, intelligence, and the multiverse existing as robust manifestations of CAP, (b) the mutability of physical constants and laws (including possibly the second law of thermodynamics) to allow more novel relations and actions to emerge over time in the universe or multiverse, and (c) the increasingly valued notions of connectedness, relationality, diversity, freedom, and complexity—possibly as universal values shared by other extraterrestrial intelligence.

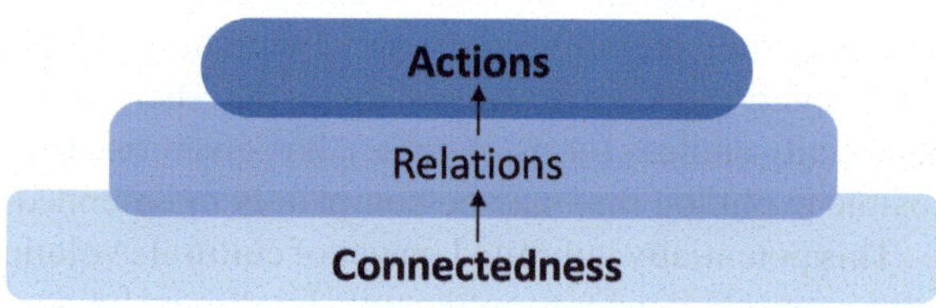

Fig. 10.2 Simplified representation of the connection-action principle

We briefly revisited the psychology of philosophy, heightening our sensitivity to the potentially aesthetic, and possibly misleading appeal of notions of connectedness and diversity, which may have evolutionary roots. That psychological sensitivity, along with more detailed practical ethical considerations, led us to raise questions about the pragmatic utility of the connection-action principle and whether value measurement in the context of CAP would be possible or credible. Nevertheless, we briefly considered at least theoretical paths forward.

We explored the possibility that there may be conceptual limitations to CAP and/or that the universe may be imperfectly, incompletely, or perhaps just differently realizing connectedness, relationality, and action compared to any present conceptions we might have for CAP and its potential implications. We considered the possibility that the universe may have to in some sense "start over" with new cycles to continue realizing connectedness, and/or that a "multiverse" or "parallel universes" may be a way to more fully realize the connection-action principle.

As a way to think about potential overlap (as shown in Fig. 10.1) or integration and synthesis between cosmocultural evolution and the connection-action principle, we can entertain the idea that as valuing cultural agents in the universe, we can still choose relations and relationality more generally, or even CAP more specifically, to be important for ourselves and the universe. As a kind of "cerebral cortex for the universe," it can be our choice—one we could make on behalf of the universe if we wish to.

10.3.4 Summary of Chapter 9: Meaning and Ethics?

The three worldviews we explored (*cosmological reverence*, *cosmocultural evolution*, and the *connection-action principle*) can be interpreted to support cosmological theories of value that ascribe various forms of value to the universe, its creations, and the role of life and intelligence in cosmic evolution. The degrees of value vary and tend to increase as we move right to left and down the relational framework in Fig. 10.1. We briefly touched on how these worldviews can ultimately ascribe value to a morally creative cosmos, one in which we emphasize respect for relationships, diversity, and perhaps complexity.

Cosmological reverence emphasizes value in the sense that the universe is valuable for beings like ourselves who can look back on our cosmic history and choose to value that cosmological story—and hence engender ethical commitments to the products of cosmic evolution and the universe as a whole. *Cosmocultural evolution* suggests that not only is the universe valuable to beings such as ourselves that have emerged from cosmic evolution, but also that we have value and significance for the universe as intentional, contemplative, cultural valuing agents who have helped bootstrapped the universe into the realms of value, meaning, and purpose. This can be seen to increase our meaning for, and ethical commitments to, the universe in the sense that we have greater responsibility to act as arbiters of value and meaning for ourselves and the universe. The *connection-action principle* (CAP) takes us quite a bit further and makes speculative claims about the fundamentally creative nature of the universe being grounded in the universe's connectedness which is manifested as relations and actions, perhaps in ever-increasing degrees—suggesting a kind of intrinsic cosmological value for things like relationality, diversity, freedom, and complexity—all of which can be thought of as robust manifestations of relations and actions.

We applied these broad "cosmological theories of value" to briefly explore examples of "cosmic ethics" and stronger versions such as "cosmocentric" ethics—specifically applied to the question of potential interactions with life on Mars. We leveraged the cosmic values of CAP such as diversity and complexity to suggest that even "primitive" Martian life is worthy of our respect and possible ethical commitment to peaceful co-existence—perhaps even more so if that life constitutes an independent origin of life which would add to and enhance the diversity of the universe. We briefly considered the possibility that the connection-action principle may also allow for some assessment or "measurement" of degrees of value, perhaps even intrinsic value, which can help inform the level of ethical commitments we might have, including when considering unusual situations like human interests in settling Mars vs. the value of indigenous Martian life-forms that might be compromised by a human presence.

10.4 Choosing a Model or Worldview

Lawrence Krauss (2017) titled his book, "*The Greatest Story Ever Told—So Far*"—a great title. The story of our universe is indeed a great open-ended story—a show of sorts for which thereis much more to come, much more to learn and share, and much more to revere as the universe continues to evolve. For some, it can be enough to simply pursue physical knowledge of the world and refrain from meaning-making and worldview-building. Scientific minimalism can indeed be its own reward. Hailing the universe's story as the "greatest ever told" is its own kind of cosmological reverence—something many have come to experience.

Exploring and building worldviews that are informed by science and modern cosmology might lead somewhere helpful (Crawford 2018a, b), but regardless, exploring worldviews in the context of science and cosmic evolution can still be rewarding in a variety of ways even if it leads nowhere in particular. So while wedo not need to choose a model or worldview if we do not think it is warranted, it may still nevertheless be fruitful to engage in a brief evaluation.

Building on previous work from other philosophers, Clement Vidal (2014) offers a complex and comprehensive metaphilosophical framework for evaluating worldviews, consisting of (a) six philosophical dimensions, (b) a philosophical agenda for defining what a worldview is, (c) nine evaluation criteria, and (d) a set of tests, including "first-order" tests of *is—ought*, *ought-act*, and *is-act*. A thorough application of Vidal's framework to the various worldviews explored in this book would be an intriguing analysis,[2] and while outcomes of such analyses would presumably vary widely depending on how one emphasizes the evaluation criteria (itself a reflection of philosophical positions that Vidal treats explicitly), there is a brief assessment that might be worth exploring. The three categories of the nine criteria Vidal lists can be loosely interpreted to suggest that bootstrapped cosmocultural evolution may best satisfy a good balance across the objective and subjective criteria, some of which will be explored briefly here., i.e., Vidal's three categories of criteria are: (1) *objective* criteria (e.g., internal consistency, consistency with science and broad scope), (2) *subjective* criteria (utility, consistency with personal experience(s), emotional value, and (3) *intersubjective* criteria (addressing social factors such as collective utility, reducing conflict, and communicability[3]).

It is notable that Vidal suggests twice as many subjective criteria than objective criteria, and this seems reasonable for evaluating worldviews that often need to address a range of subjective interests of individuals and groups. However, it would also be reasonable to emphasize objective criteria, in which case an evaluation assessment might come out quite differently. Regardless, any evaluation would likely come out differently depending on any number of nuances and biases associated with almost all of the nine criteria. Related to those potential nuances, the objective criteria could also include an additional, or related criterion, namely something like "degree of speculation" or "metaphysical content" or "degree of metaphysical speculation," which will have relevance to the following brief worldview comparison.

Scientific minimalism, as characterized in this book, is a safe and prudent worldview in the sense that most versions of it acknowledge our cosmological relationship(s), but in general it does not infer significant value from those relationships—it does not bridge facts and values and commit the "naturalistic fallacy" (although stronger versions of scientific minimalism can be interpreted to overlap with cosmological reverence as shown in Fig. 1.1). Scientific minimalism can suffice for many people and perhaps increasingly in the longer term as humanity becomes more at ease with what may ultimately be an objectively pointless universe (Lupisella 2019).

For others, cosmological reverence may be a welcome step in going from a minimalist scientific cosmological worldview to valuing the universe in varying degrees as a source of origins and ongoing intimate relationality. Many scientists and laypersons may see themselves in this overlap between scientific minimalism and cosmological reverence. Scientific minimalism can be seen as not saying much about what we should aspire to, but it at least acknowledges our apparent relationships to the universe and our environment more generally and so may promote a kind of basic pragmatic respect for the broader world, without having any particularly strong philosophical or ideological commitments.

---

[2]An initial attempt (Lupisella 2019) at a more comprehensive comparison reveals a number of complexities and challenges that would benefit from a more detailed analysis, or ideally, analyses by multiple evaluators.

[3]Vidal refers to communicability more narrowly as "narrativity," but I am choosing to use what I see as the slightly broader term of "communicability."

While cosmological reverence is notably different from scientific minimalism, it is not pantheism. Cosmological reverence simply makes the choice to value the universe for scientific and "relational" reasons: we arose from the universe and are intimately bound up in it and related to it, but we do not have any particular broader cosmic significance. This may not be enough for certain purposes or for many people more generally. Cosmological reverence does not compel highly specific norms or actions, or long-term meaning or purpose or any kind of direction or teleology in the universe. It is instead a kind of passive reverence for the cosmos that can nevertheless inspire and inform certain ethical questions such as how we might value certain cosmic creations, including other life-forms.

Cosmocultural evolution, however, is a stronger view in the sense that it can suggest a certain amount of cosmic responsibility for intelligent beings because cultural evolution has the potential to have high degrees of significance for the cosmos as a whole—but again, for no other reason other than cultural agents arose via physical processes of cosmic evolution (or, as noted prior, what might be thought of as "bootstrapped cosmocultural evolution") and now have agency and can choose and act on a kind of cosmocultural evolution value system or worldview. We saw an earlier quote from Paul Davies that "Somehow, the universe has engineered not only its own self-awareness, but its own self-comprehension." (Davies 2009, 385). As noted previously, we might add that the universe has "engineered" its own *self-valuing*.

The connection-action principle goes much further and contemplates values based on a suggested "objective" (essentially "metaphysical") nature of the universe. The connection-action principle would need much more detailed investigation to better explore any ultimate truth-value it may have, and it is potentially intellectually and philosophically problematic for some of the reasons noted previously. While it is arguably supported by a number of lines of reasoning and some consistency with scientific and philosophical ideas explored previously, it is nevertheless primarily a metaphysical set of claims that arguably lack sufficient physical commensurate evidence or sufficient predictions and tests needed to be highly persuasive. Its potentially "universal" and prescriptive moral and ethical implications are not likely to be something many philosophers or laypersons could easily defend or identify with. Manifesting the connectedness of the universe through myriad forms of relations and actions may not ring true for many, and of course it may not be true.

Of the views developed in this book, this leaves cosmocultural evolution, particularly *bootstrapped cosmocultural evolution*, with its implications for a morally creative cosmos that cultural agents can help create, as perhaps the most tenable "meaningful" worldview. While this kind of worldview construction could be another example in a long list of speculative attempts to find meaning and purpose where there otherwise is none, it does appear to be defensible. It appears we can indeed say that complex and robust forms of value, meaning, and purpose do exist in the universe at least in the form of human minds—and perhaps in the minds of other creatures as well in various forms and to varying degrees.

This realm of value, meaning, and purpose has then literally emerged in the universe through cosmic evolution, through the activities of our minds and perhaps other minds that may exist throughout the universe. The potential for this valuing capacity and meaning-making may have unlimited potential for the universe as a whole. There may, of course, be other forms of value that are independent of beings like ourselves (as suggested by the connection-action principle), but it does nevertheless appear that beings like us are at least one means by which the universe is finding, or "discovering", or bootstrapping forms of value, meaning, and purpose in what may be an extremely large, if not infinite, possibility space of such pursuits.

The claim that value, meaning, and purpose have emerged in the universe as a product of cosmic evolution is in some sense a minimalist claim and perhaps even a trivial claim to some, but it is potentially consequential nonetheless. It is intellectually and philosophically minimalist in the sense that there is no need to invoke some other kind of dynamic or force or substance in the world such as "spirit" or God. There is no appeal to transcendent realities or unusual concepts or constructs. However, the implications and significance are still notable in the sense that if value, meaning, and purpose have emerged in the universe through us, then we are arguably responsible for its realization.

We are co-creators and arbiters of value that can make the universe valuable in a general sense, and also pursue very specific forms of value, e.g., having to do with morality and/or diversity and/or endless morally creative endeavors. Indeed, if we choose, we can perhaps make the universe "teleological" in the sense of choosing purposeful directions or trends for the universe—including its eternal existence if that is deemed to be worthy of our "normative aspiration." It is up to us to decide, to choose. Presumably, there will be many such pursuits, highly divergent and often in conflict, which call for careful and measured deliberation and meaning-making.

As touched on previously, one possible way to loosely integrate bootstrapped cosmocultural evolution and the connection-action principle is to consider the possibility that intelligent beings, via their cosmocultural evolution, could simply choose to emphasize something like a connection-action principle (e.g., particularly softer versions noted previously—hence the small overlap between cosmocultural evolution and connection-action principle shown in Figs. 1.1 and 10.1). This could simply be a philosophical choice, where no metaphysical claim is necessarily made about its objective truth-value, but is instead emphasized for any number of other reasons—including aesthetic appeal or any practical utility it may have for individuals or groups.

If the connection-action principle is not independently true of intelligent valuing agents, it may nevertheless have sufficient appeal for other reasons, and if we value it, we may see ourselves and the universe as benefitting from more connectedness and relations and actions on behalf of those connections—including our relationships and actions toward other beings and the broader world. As creators and arbiters of value in the universe via cosmocultural evolution, we can, at least in some general sense, make the connection-action principle "true enough" for ourselves and the universe as part of the bootstrapped cosmocultural evolutionary process we can play a critical role in.

So, in the spirit of a kind of "Goldilocks" assessment, perhaps cosmological reverence is in some sense "too cold," suggesting there is not much, if any, cosmic significance for intelligence or cultural evolution; the connection-action principle is "too hot," making strong metaphysical claims and implying value-laden "objective" metaethical claims that are difficult to defend; and bootstrapped cosmocultural evolution is "just right"—bridging what may be pointless humble beginnings to cosmically meaningful futures.

Regardless of whether life and culture arose by chance or as part of some "cosmic imperative," the value of life, intelligence, culture, and cosmic creations can still be appreciated. Regardless of whether we are a rare cosmic gem, part of a cosmic fugue, or part of some deeper cosmic meaning or purpose, there is some kind significance we can claim for life, mind, and culture. We can legitimately see ourselves as meaningful and purposeful, worth preserving and developing—for ourselves and the universe. Let us then play worthy roles in *The Cosmos*, the greatest show we know.

References

Crawford, I. A. (2018a). Widening perspectives: the intellectual and social benefits of astrobiology (regardless of whether extraterrestrial life is discovered or not). *International Journal of Astrobiology, 17*(1), 57–60.

Crawford, I. (2018b). Big history and the cosmic perspective. *Astronomy and Geophysics, 59*, 5.33–5.36.

Davies, P. (2009). Life, mind, and culture as fundamental properties of the universe. In S. J. Dick & M. Lupisella (Eds.) *Cosmos and culture: Cultural evolution in a cosmic context*. Washington, DC: NASA History Series. http://history.nasa.gov/SP-4802.pdf.

Haidt, J. (2012). *The righteous mind*. New York: Random.

Krauss, L. (2017). *The greatest story ever told—so far. Why are we here?* New York: Atria Books.

Lupisella, M. (2019). Is the universe enough? Can it suffice as a basis for worldviews? *Journal of Big History, III, 3,* 123–140.

Vidal, C. (2014). *The beginning and the end: The meaning of life in a cosmological perspective*. Berlin: Springer.

Author Index

M. Lupisella, *Cosmological Theories of Value*, Space and Society,
https://doi.org/10.1007/978-3-030-25339-4

Subject Index

M. Lupisella, *Cosmological Theories of Value*, Space and Society,
https://doi.org/10.1007/978-3-030-25339-4

The manufacturer's authorised representative in the EU is Springer Nature Customer Service Centre GmbH, Europaplatz 3, 69115 Heidelberg, Germany. If you have any concerns regarding our products, please contact ProductSafety@springernature.com

Printed and bound by CPI Group (UK) Ltd, Croydon, CR0 4YY
22/09/2026
02221057-0003